Bonanza Rich

Bonanza Rich

Lifestyles of the Western Mining Entrepreneurs

Richard H. Peterson

University of Idaho Press
Moscow, Idaho

University of Idaho Press
Moscow, Idaho 83843

Design by Karla Fromm
Printed in the United States of America
96 95 94 93 92 91 5 4 3 2 1

Library of Congress Cataloging-in-Publication Data

Peterson, Richard H.
Bonanza rich: lifestyles of the western mining entrepreneurs / Richard H. Peterson.
p. cm.
Includes bibliographical references and index.
ISBN 0-89301-143-6
1. Mineral industries—West (U.S.)—History—19th century. 2. Industrialists—West (U.S.)—History—19th century. 3. Millionaires—West (U.S.)—History—19th century. 4. Wealth--West (U.S.)—History—19th century. 5. West (U.S.)—Social conditions. I. Title.
HD9506.U63A17155 1991
338.7′622′092278—dc20
[B] 90-25792
CIP

To
Nora and Nina
and
the memory of my grandmother,
Mrs. Martha C. John

Contents

Preface

THE GENERAL PUBLIC CONTINUES TO BE fascinated with the romance and glamour of the western mining frontier. The frantic rushes to new discoveries, the legends of lost mines, the colorful characters—from prospectors to prostitutes—that populated the mining camps, and the elusive dream of sudden success still excite the imagination. However, since World War II, scholarly study of western mining has concentrated on the mineral industry itself. The noted mining historian Clark C. Spence observes that "historians of the mining West have moved away from the romance and narrowness of the early frontier years to put more emphasis on the industry, the complexities of economic and technological change, and the reaction of workers and communities to diverse parts of the transition."[1] My earlier work on the subject, *The Bonanza Kings: The Social Origins and Business Behavior of Western Mining Entrepreneurs, 1870–1900* (University of Nebraska Press, 1977) is part of this shift in historiographical emphasis. Based upon the careers of fifty prominent owners of mines, mills, and/or smelters in the precious and related nonprecious metals mining industry of the trans-Mississippi West, the book examines how individuals succeeded in the industry by securing and managing such factors of production as property, labor, capital, and technology.

This sequel examines the consequences—rather than the causes—of the success experienced by these same fifty mining entrepreneurs. It discusses and analyzes the ideology, political activities, philanthropy, conspicuous consumption, cultural interests, recreational roles, and family relations of the mining men during the late nineteenth century. E. L. Godkin, editor of the *Nation*, wrote of this period and its new rich that the United States is a "gaudy stream of bespangled, belaced, and beruffled barbarians. . . . Who knows how to be rich in America? Plenty of people know how to get money; but . . . to be rich properly is, indeed, a fine art. It requires culture, imagination, and character."[2] This study seeks to explain how the mining leaders attempted "to be rich," by addressing several questions: How did the mining moguls explain and define their success in business given the popular beliefs and values of late–nineteenth century America? What kind of lifestyle was made possible during the Gilded Age by the acquisition of mining fortunes? Why did the mining capitalists choose to spend their wealth as they did?

The significance of the study rests in large part with the attempt to relate the western experience to that of the nation. Basing their assumptions on the writings of Frederick Jackson Turner, the distinguished frontier historian and theorist, historians of the West traditionally have viewed the area as unique or culturally distinctive, sometimes implying that it evolved almost in a vacuum. It is my contention that the western region can best be understood as part of the nation. Indeed, an examination of the fifty bonanza kings provides a convenient bridge between the East and the West because western mining leaders and eastern business elites largely shared a common lifestyle.

In order to adequately represent the general leadership of the western mining industry, the subjects of this study were selected on the basis of financial success and several other criteria. Thirty-six of them or members of their immediate families were regarded as millionaires according to surveys conducted by the *New York Tribune* in 1892 and the *New York World* in 1902.[3] Undoubtedly, many held high, if not the highest, positions in the largest mining companies as defined by capitalization or production. Their business leadership was comparable to that of those who dominated the fields of steel, oil, and railroads.

Most of the mining entrepreneurs also were chosen because they appeared in the *Dictionary of American Biography* or the *National Cyclopedia of American Biography*, each of which listed the country's most prominent or successful business leaders. In a collective study of 1,464 such executives born between 1570 and 1879, C. Wright Mills made his selections from the *Dictionary of American Biography*, reasoning that it "forms a convenient point of departure for an over-all view of the social characteristics of eminent American businessmen."[4] A scholar who has developed a specialized strategy for collecting data on American business officials calls the *National Cyclopedia of American Biography* as complete a source of systematic information on businessmen of the late nineteenth century as any such compendium."[5] Of the fifty mining magnates, thirty-two were the subjects of biographical essays in one of these reference works, and some appeared in both.

Nonabsentee ownership provided an additional criterion for selection. No attempt was made to eliminate those who became absentee owners after mining properties had matured. Nonetheless, most of the men chosen had joined the early rushes to the major mining frontiers and districts and had stayed on to mature with their properties. Their early arrival in the West helps explain their success. These westerners can be contrasted with such eastern mining financiers as Henry H. Rogers of the Amalgamated Copper Company or Daniel Guggenheim of the

American Smelting and Refining Company, who operated their enterprises from a distance and usually invested in western mining after it had become heavily industrialized.

In the interest of balanced regional representation, an attempt was made to select moguls from nearly all of the western mining frontiers and the most productive mining districts. Although these bonanza kings are identified with their major fields of investment or operation in the appendix, they are not differentiated specifically by region in the other tables. In view of the multiple regional investments of most mining entrepreneurs in this study, such a differentiation would be misleading if not somewhat unrepresentative.

Consideration of time as well as place helped to determine the sample. The period from 1870 to 1900 was singled out for special attention. Despite the continuing instability of frontier social and economic conditions, this era saw western mining mature into a technologically sophisticated industry surrounded by an increasingly stable population base and labor force. In particular, the wandering prospector had given way to the resident wage-earning miner. On the whole, the subjects of the book were most productive and most prominent during these years, although their mining careers often had begun before 1870 and in some cases continued beyond the turn of the century. Indications of career length and success can be derived roughly from the birth and death dates available in the appendix.

Around 1900 the industry experienced structural changes in the form of increased corporate consolidation and eastern financial domination. Even before 1900, mining was becoming a part of the national economy rather than continuing as an isolated western endeavor.[6] Such heavily capitalized national corporations as the American Smelting and Refining Company, Phelps Dodge Corporation, and the Amalgamated Copper Company were beginning to dominate the industry as completely as the western bonanza kings previously had controlled local mining districts.

The turn of the century thus seems appropriate as a terminal date for the selection of leading western mining careers.

From a practical standpoint, readily available biographical information also influenced selection of the sample.[7] Although the book is intended largely as a synthesis of secondary sources, unpublished primary sources were not neglected. Personal and business papers were utilized for some of the entrepreneurs and their wives. The Thomas F. Walsh Family Papers in the Library of Congress, the Jesse Knight Family Papers in the Brigham Young University Archives, the Horace A. W. Tabor and Nathaniel P. Hill Papers in the State Historical Society of Colorado, the John P. Jones Family Papers in the Special Collections Department of the University of California at Los Angeles, the Phoebe A. Hearst Papers in the Bancroft Library on the Berkeley campus of the University of California, and the Simeon G. Reed Papers at Reed College were typical of those consulted. Unpublished biographical theses and dissertations were found for eight of the men, and twenty-four entrepreneurs were the subjects of published biographies, autobiographies, memoirs, and magazine articles. In sum, for the purpose of this study a mining elite or bonanza king is generally defined as a wealthy, western owner of various late nineteenth-century mining and related enterprises, who was important enough to be included in the contemporary biographical encyclopedias or to be remembered by modern historians.

In researching the lifestyles of these fifty men, I have benefitted substantially from the resources and cooperation of many individuals and institutions. Special thanks must be extended to the courteous staffs of the Inter-library Loan Department at San Diego State University and the University of California at Davis, the Bancroft Library, the California Historical Society, the California State Library and its Sutro Branch in San Francisco, the San Mateo County Historical Association in California, the Reed College Library, the Utah State Historical Society, the Brigham

Young University Archives, the State Historical Society of Colorado, the National Archives, and the Manuscript Division of the Library of Congress. I am indebted to Professor Norris Hundley of the University of California at Los Angeles and the able editorial referees of the *Pacific Historical Review*, who helped me refine the argument in the second chapter. A faculty summer fellowship from San Diego State University allowed me to extend my research for the study. Finally, I would like to express a very personal note of gratitude to those who gave up so much to make this book possible, my wife Nora and our daughter Nina.

NOTES

1. Clark C. Spence, "Western Mining," in *Historians and the American West*, ed. Michael Malone (Lincoln: University of Nebraska Press, 1983), 109.

2. As quoted in Carl N. Degler, et al., *The Democratic Experience: An American History*, 5th ed., vol. 2 (Glenview, Ill.: Scott, Foresman and Co., 1981), 49.

3. See Sidney Ratner, ed., *New Light on the History of Great American Fortunes, American Millionaires of 1892 and 1902* (New York: Augustus M. Kelley, Inc., 1953). The 1892 survey listed the occupations in which the fortunes were made.

4. C. Wright Mills, "The American Business Elite: A Collective Portrait," *Journal of Economic History* (special issue *The Tasks of Economic History*) 5 (December 1945): 20.

5. William G. Roy, "Collecting Data on American Business Officials in the Late Nineteenth and Early Twentieth Century," *Historical Methods* 15 (Fall 1982): 144.

6. Rodman W. Paul, *Mining Frontiers of the Far West, 1848–1880* (New York: Holt, Rinehart, and Winston, 1963), 194.

7. The availability of adequate biographical information in part prompted Frances Gregory and Irene Neu to select their sample of industrial leaders from the decade 1870-79. See Frances W. Gregory and Irene D. Neu, "The American Industrial Elite in the 1870s, Their Social Origins," in *Men in Business: Essays in the History of Entrepreneurship*, ed. William Miller (Cambridge, Mass.: Harvard University Press, 1952), 195.

1

The Ideology of Business Success

IN THE LATE NINETEENTH century the wealthy in America were legitimized by a set of ideas that attributed their success to hard work, self-denial, rugged individualism, natural selection, and even God's will. According to such familiar Horatio Alger stories of the era as *Work and Win*, *Strive and Succeed*, and *Risen from the Ranks*, poor boys could grow up to be successful, self-made men if they were industrious, thrifty, and temperate in character. Often overlooked in these melodramatic novels was the fact that poor ambitious youths who did succeed frequently had been helped by chance meetings with wealthy benefactors who gave them the break or head start their good character or good deeds warranted. According to the popular social Darwinist philosophy of the period, which applied Charles Darwin's theories of biological evolution to society, those who succeeded were naturally selected by virtue of being able to compete effectively with their adversaries and adapt to their social and economic environment. The contradiction between the free will of Alger's heroes and the inexorable laws of nature inherent in social Darwinism apparently did not trouble American industrial leaders, if they gave it any thought at all. They were the fittest who won the competition for wealth because God or Nature had endowed them with superior talents, character, and rugged individualism. In theory, government was supposed to encourage this economic development by adopting a laissez-faire philosophy which would not interfere in the process of free competition and natural selection. In practice, business leaders insisted that government promote and support industrial expansion, but not stifle it through regulation. The concepts of rugged individualism, natural selection, and laissez-faire economics comprised the ideological core of capitalism in America's Gilded Age. With this in mind, how did the mining leaders explain or defend their success? In their view, what personal character traits or values led to achievement in business? How did they perceive of the proper role of government in the economy? In short, how did they define the ideology of success in the business civilization of America in the late nineteenth century?

Some of the bonanza kings recorded their thoughts on how to succeed in business, and their recommendations testify to the alleged importance of the familiar Alger-style virtues in achieving success. As an eminently prosperous entrepreneur, Thomas Walsh frequently was asked to comment on his rise to fortune. He had been a longtime luckless prospector until he unlocked the golden treasure of the Camp Bird mine in Colorado's San Juan District in 1896. The mine yielded $2.5 million before Walsh sold out in 1902 for over $5 million.[1] Like other contemporary business leaders, Walsh attributed much of his success to hard work: "My position in life has come from hard labor allied to my natural ability in discovering and testing the precious metals."[2] His dedication to systematic, diligent labor is perhaps best expressed by his daughter Evalyn's observation that he was not inclined to follow the example of "the careless prospectors who ceased to be . . . methodical in their search; who were always ready to assume that the first prospectors to look at any piece of ground had been thorough."[3] Walsh also noted the value of good character, thrift, and honesty in achieving success: "Without having many and upright principles you cannot be a factor in life, nor can you reach the goal of real success. Character is one of the greatest powers in the world. . . . Be frugal and saving. Honest wealth is a badge of power; it makes possible the pleasure of doing good."[4] Finally, in an address to a school graduating class he emphasized the rewards of self-reliance: "You are mariners about to take a voyage over untried seas. Your destination is the harbor of honorable success. To gain this much prized harbor, each one must sail his bark alone, must be captain, pilot and crew in one, and depend upon himself only."[5] Walsh's commitment to such values may explain why his son's boyhood room was equipped with the works of Horatio Alger, although they seemed peculiarly unnecessary for the child of such a wealthy man.[6]

Jesse Knight, the Mormon mining mogul whose development in the 1890s of the Humbug Mine in Utah's Tintic District laid the foundation for his substantial fortune, shared Walsh's belief in the cult of the self-made man. Like his Colorado counterpart, Knight saw industry and frugality as critical to success in business judging by the advice he gave a young man seeking employment: "The best advise [*sic*] that I can give you and that which I think have been [*sic*] more valuable to young men than either giving or lending them money has been . . . for them to find work and save their earnings and in this way enter into whatever they chose to do either in a line of business or in obtaining an education. The men who have acompolished [*sic*] the most are those who have had to work at 'Manual Labor.' That is the way I got my start in life and without the desire to work one's chances in this life are not the best for success, and no matter what one makes a success of thers [*sic*] always plenty of work to do to make that success continue."[7]

Horace Tabor, whose development of the Little Pittsburg and Matchless mines in Leadville, Colorado, quickly made him a multimillionaire, similarly identified success with making the most of opportunity by being hard-working and thrifty. Reflecting upon his youth in New England, he drew a simple but fairly typical picture of how to prepare for prosperity: "In regard to success in life, I think that is to a great extent opportunity. We are all creatures of accident more than anything else. I do not know as there was much accident in my case. . . . At school I was always well near the head of my class. I have been very industrious all of my life. . . . In the district in which I was brought up in Vermont. . . one of the first things necessary was that a person should be of economical spirit. He was compelled to be; it was a necessity. I always hated to see anything go to waste."[8] Like Tabor, Joseph R. DeLamar, who developed mines in Colorado, Idaho, and Utah, associated success with exploiting opportunity: "Every human being, no matter how he may be situated has a certain number of chances in his lifetime, which, when properly utilized, lead to success and prosperity. But very few people realize the importance of utilizing their opportunities promptly."[9] According to Irving Howbert, who laid the basis for his fortune as a part-owner of the Robert E. Lee silver mine at Leadville in 1878, early disadvantage in life would properly equip an individual to seize opportunity: "In this Western country there are singularly few men who have accomplished great success for themselves, who have been swathed during the formative period of their lives in the enervating folds of luxury."[10] In short, hard knocks and hard work would eventually pay rich dividends. Such thinking was not surprising since "of the cultivated qualities that helped men to rise, industry was the one most often prescribed and most elaborately justified by self-help advisers."[11]

William C. Ralston and Simeon Wenban would have agreed that those who worked hard would be duly rewarded. Ralston, cofounder in 1864 of the Bank of California in San Francisco and a financial developer of extensive properties on Nevada's Comstock Lode, pointed out: "I was brought up to hard work in various mechanical branches from my boyhood, when laborers were required to toil from 12 to 15 hours per day and I still find it often necessary to devote as many hours to my present vocation."[12] Wenban, the owner of silver mines in eastern Nevada, implied that long hours of labor as a youth on the family farm in Ohio had prepared him for comparable work in the California and Nevada mines: "Being always active and restless I. . . turned my attention to work, and worked so hard, that I did not have time to eat. As an illustration I. . . would go out [to the fields] in the morning after breakfast and the chances are I would not be home again before ten o'clock at night."[13] Later, on the California mining frontier in the 1850s, he reportedly labored as long as twenty hours a day to liquidate debts arising from conflicting titles to placer mining claims.[14]

For Dennis Sheedy, the cattleman who had become one of Colorado's major smelting and refining entrepreneurs by the turn of the century, the keys to success were not only hard work but self-reliance, opportunity, and the right kind of character. Reminiscing on his long career, he stated: "My story is of a boy of sixteen who started out alone in the world to carve out a fortune, and did, includ[ing] an active participation in the making of the West and of the Southwest." Although Sheedy may have had an inflated sense of self-importance, like other mining men he equated success with the ability to maximize opportunity: "I look around and I see in all lines of business young men who in the years to come will have grasped opportunity and will be the captains of industry, as I see men today whom I knew when they were starting out in life, determined to succeed, and did."[15] He also attributed his success to the values inculcated in him as a child, recollecting that he received from his parents a "very strong sense of honesty and integrity, and of the golden rule. . . . It is to this teaching that I owe all that I have been, and am."[16] He acknowledged that his parents taught him the importance of thrift: "My father and mother were quite insistent that we children earn our spending money, and thereby they impressed upon us the necessity of taking care of ourselves, and of saving." Of the industry and determination which allegedly made for success by popular nineteenth-century standards, he said that his first real ambition was "an ambition to be first... throughout my life there was that ambition—to be first, to succeed in whatever I attempted."[17] As envisaged by Sheedy, the path to prosperity required not only certain character traits but an achieving attitude, a will to succeed.

George Hearst, who either owned or had an interest in such fabulously productive mines as the Ophir in Nevada, the Ontario in Utah, the Homestake in South Dakota, and the Anaconda in Montana, shared Sheedy's sentiments about the importance of ambition. Although he had very little formal education, he recalled: "I liked school firstrate and was very ambitious to learn everything, and it would worry me terribly if I could not succeed."[18] As a youth, he also developed a stubborn individualism that characterized his later career: "I never did a sum [arithmetic] in the way the rule stated. I would always figure it out in my own way, I always had to have it my own way or not at all, and this has been my disposition all through life."[19] Finally, during Hearst's childhood on the family farm in Missouri, he acquired a spirit of self-sufficient enterprise that undoubtedly helped him later in business: "We made our own clothing and used to beat the flax and cottonseed out on the hearth before the fire, and you may be sure that gave us plenty to do. . . . As soon as I could do anything I would try to make myself useful. Everybody had to be busy, as we had to make everything. In fact I never remember a time when there was not something to do. We had three or four slaves but had to work very hard ourselves all the same. I was my mother's boy, looked after the ducks and chickens and gathered up the eggs until I got big enough to cut wood."[20] Thus, based upon the social conditioning of his youth, Hearst must have regarded ambition, individualism, and hard work as essential prerequisites for a successful career.

William A. Clark, a flamboyant Montana mining mogul, built the first smelter in Butte for silver and copper and owned the United Verde Mine and smelter in Arizona. When asked to offer guidelines to ambitious youth, he focused on the need to develop a high moral character: "The most essential elements of success in life are—a purpose, increasing industry, temperate habits, scrupulous regard for one's word, faithful performance of every promise, perfect system in business so as to be in close touch with all details, putting nothing off until tomorrow, courteous manners, a generous regard for the rights of others, and, above all, integrity which admits of no qualification or variation."[21] Although Clark did not always honor the more saintly of these suggestions, especially in his notorious political struggle with Montana copper mining rival Marcus Daly, he obviously saw business achievement as a reflection of personal character.[22] His most recent biographer regards Clark's career as a "striking illustration of the Protestant 'work ethic' if ever there was one."[23]

John Mackay, one of the Comstock Lode's so-called "Big Four," emphasized self-reliance as the key to outstanding achievement in business. When confronted with striking miners at one of his Virginia City, Nevada, properties, he expressed a philosophy of self-sacrificing rugged individualism that was incompatible with the strike: "The little additional money that these miners want is nothing. What I hate is the spirit of it all. I rolled rocks in the Yuba River month after month, even though I did not earn four bits a day, but then I did not strike. I lived on the four bits until I could make more, then I enlarged my menu, and the one thought that possessed me in all those years was, sometime, somewhere, if I had but courage enough and strength enough, I could win out. I never thought of asking help of any man, I never growled at conditions; the good God had given me a good constitution and a pair of strong arms, and I always said to myself that that was capital enough to begin with in this world."[24]

The commitment of some mining leaders to self-reliance or rugged individualism is suggested not only by what they said about themselves but by how they reacted to others. Such was the case in the depression year of 1894 when Jacob S. Coxey's army of the unemployed marched in protest on Washington, D.C., and inspired as many as seventeen similar "industrial armies" throughout the country to start for the nation's capital.[25] This brought a strong reaction from John Q. Packard, a forty-niner and later the owner of such profitable mines as the Tiptop and Michigan in Idaho and especially the Eureka Hill in Utah. According to a contemporary observer of mining elites, "he was wonderfully wrought up when the Coxie army started on its march. He had never relied upon anyone save himself and could not comprehend how any healthy man should directly or indirectly beg in a land like ours. For a long time he had been noiselessly contributing to maintain a certain church in Salt Lake City. No one knew it but the pastor of the church. As the army neared Salt Lake, just as Packard was most furious about it, he met this pastor on the street, who greeted him with, 'Oh, Mr. Packard, what can we do for these poor men? Where can they go?' At the top of his voice Packard shouted: 'Let them go to h—l!' And strode on leaving the good pastor paralyzed with astonishment."[26] In effect, Packard seemed to be saying that America, even in the throes of a deep depression, was still a land of opportunity that would reward those who were willing to depend on themselves rather than demand government work relief programs.

Since not all bonanza kings expressed personal opinions about the nature of success, it is necessary to look elsewhere to find further evidence of how as a group they conceived of achievement in business. Contemporary biographical encyclopedias and so-called "mug books" popular in the Gilded Age give insight into the character traits leading businessmen associated with success. Often the subjects of these works authorized and subsidized their own biographies, so they undoubtedly agreed with what was written about them. For example, over 2,500 Montanans paid the publisher A. W. Bowen and Company of Chicago to include their biographies in the massive 1902 edition of *The Progressive Men of the State of Montana*.[27] In a comparable contemporary biographical reference, Simeon G. Reed, the Portland steamboat transportation magnate who also owned the famous Bunker Hill and Sullivan silver-lead mine in northern Idaho's Coeur d'Alene district from 1887 until 1892, was described as "full of resources and. . . strongly self-reliant and. . . not afraid to stand alone."[28] It may be assumed that Reed attributed his success in part to this kind of rugged individualism. Likewise, Winfield S. Stratton, whose discovery of the Independence Mine at Cripple Creek, Colorado, in 1891 turned a struggling prospector into a millionaire, must have accepted the favorable assessment of a contemporary biographical source that described him as having "the air of independence and self-reliance that belongs to a strong man of substantial character and latent force."[29] The same source called Irving Howbert "a splendid type of the successful man. He stands for the best things in life—honesty, integrity of character, self-reliance, educational [*sic*], morality and Christianity."[30] An example of the popular "success literature" of the period left this description of Henry R. Wolcott, the general manager of the Boston and Colorado Smelting Company in the 1880s and an owner of mines in such districts as Leadville and Creede, Colorado: "He worked his way up from the ranks by hard work and merit and his career is not only an object lesson but an inspiration for the young men of today."[31] Wolcott himself claimed: "I left school in the first place because I wanted to, I had been sick and was behind in my class, and father was very poor, and he was willing that I should, and I got my own place, and I have sustained myself and a good deal more since that time."[32] He reported in a letter to his brother, dated 21 December 1869, that he "reached Colorado with thirty-five cents in pocket and not even an acquaintance."[33] David H. Moffat, a pioneer Denver railroad builder, president of the First National Bank of Denver, and a successful mining investor, also had gone to Colorado as a young man to make his fortune. A contemporary business journal probably captured his own thoughts when it described him as a self-made man who had "started in life as a poor boy and amassed a large fortune entirely through his own efforts."[34] Moffat's biographer contends that "he was ambitious as a youngster and as he grew so did his view of the possible. Moffat was a typical restless young man during America's middle years, always on the prowl for new opportunity and adventure."[35]

The notion that a poor boy with a limited formal education could make something of himself by dint of ambition, hard work, and sacrifice was given credence by the following contemporary portrait of Marcus Daly of Anaconda mining fame: "His life contains great lessons to young men beginning a career of honorable ambition,—none more impressive than this: That in our favored land industry and talents will overcome all obstructions. It is difficult to recall any other American whose career proves this fact more clearly than his. He entered upon life's ordeal with nothing to rely on but his own active and resourceful mind and his indomitable will. The position from which he started to achieve all his success was humble and unpromising. His way was beset with difficulties and disappointments. Yet over all obstacles he triumphed gloriously."[36] Assuming that Daly approved of this laudatory character sketch, he seemed to be saying that poverty was the best training ground for success because it produced the desire and values necessary for self-advancement. Hearst made the same point when he wrote that "rich people are far more likely to raise poor children than poor people are. . . . The strong men of the country are bound to come from among the workers of the country. They not only raise children to be industrious, but they have to do it."[37]

Based upon what the bonanza kings wrote about success and what was written about them, outstanding achievement in business would come to young men who were opportunistic, ambitious, industrious, frugal, honest, self-reliant, and self-sacrificing. In other words, if they adopted the familiar virtues of the heroes in the Horatio Alger stories, they could make a difference in their careers and lives. It is not surprising that so many mining entrepreneurs apparently believed in this scenario for success. As a group, they had come to positions of affluence and influence predominantly from lower- and middle-class origins. The class identifications of the western mining leaders are based upon their level of formal education and their fathers' occupations and general economic standing. For example, a father designated as a "farmer" usually owned a small family-operated farm and had an income and status more nearly like that of the lower class than the middle class. In the criteria developed by William Miller in his pioneering study of the American business elite from 1901 to 1910, wealth or political eminence, often associated with the business executive or professional, defines the upper class.[38] James B. Haggin, whose partnership with Hearst yielded success in various mining fields, was born into this group. Haggin's father, a wealthy Kentucky lawyer, was able to give his son a college education and postgraduate legal training.[39] Again using Miller's criteria, those in between—including some members of the business community with no special claims to wealth or power or such professionals as the average cleric or lawyer—are regarded as members of the middle class. Besides occupational categories, the ability of the parents to give their son an education was considered in computing class origins, especially when the economic standing of the parents was unknown. A common or grammar school education helped to identify the lower class, while high school or college education gave evidence of an upper-class background. The terms "class" and "status" should be used cautiously here, since precise definitions are not possible.[40] Of forty-five western mining leaders active between 1870 and 1900, twenty are considered to have come from upper-class backgrounds, forty-four from middle-class, and thirty-six from lower-class.

The rate of social ascent on the mining frontier becomes more dramatic when one looks beyond the superficial percentages into the personal backgrounds of individual mining leaders. Some renounced the assets of a prosperous, well-connected family for the opportunity to make it on their own. Moffat, classified among the upper class on the basis of his family's economic and political standing, went to work when only twelve years old after completing common school, even though his father could have given him a college education.[41] Other bonanza kings, ranked in the middle class because of their fathers' occupations, were forced into economic independence at an early age by the untimely illness or death of the family provider. Joseph Walker, who with his younger brother Matthew became prominent in Utah merchandising, banking, and mining, took his first job after the terminal illness of his father and the loss of the modest family fortune.[42] Captain Joseph R. DeLamar overcame a similar handicap. His father, a relatively prosperous Dutch bank manager, died when his son was only six years old, forcing the latter to seek a life at sea when just a boy.[43] Thus, the mining magnates had even more reason to believe in the cult of the self-made individual than is suggested by the high percentage of men from lower- and middle-class backgrounds who rose in status. Their own careers seemed to demonstrate that those born at the bottom or in the middle of the social ladder could climb to the top if they applied the attitudes and character traits that comprised the popular ideology of business success.

Only a few bonanza kings gave much indication of their opinion of another component in the ideology of success, namely social Darwinism. Hearst apparently endorsed the philosophy when he wrote: "I do not know much about books;. . . but I have traveled a good deal and have observed men and things and I have made up my mind after my experiences that the members of the Senate are the survivors of the fittest."[44] He undoubtedly was referring to himself because he, like other business leaders, served in the U.S. Senate. Hearst may also have been echoing the familiar concept of economic conflict and competition in social Darwinism when he remarked that "the industrious man will always come to the front and the others will drop back and disappear."[45] According to social Darwinism, society was a jungle in which the strongest or fittest inevitably triumphed over the weaker members who fell by the wayside. This was part of the wholesome weeding-out process that improved the human race and allowed society to progress.[46]

Simeon G. Reed and Winfield S. Stratton must have shared Hearst's views on the subject. A contemporary biographical sketch prepared for Hubert H. Bancroft's *Chronicles of the Builders of the Commonwealth* claimed that Reed's successful efforts to dominate the steamboat traffic on the Columbia River "raised the question of the survival of the fittest."[47] This interpretation was based on Reed's interview with a member of Bancroft's staff and no doubt met with his approval. More convincing evidence of his personal beliefs can be found in his decision in 1891 to support the John Fiske Lectures with a contribution of one hundred dollars.[48] In addition to being remembered as the historian who applied the label "critical period" to the Confederation era in the nation's early history, Fiske has been called "the most persuasive of the social Darwinists."[49] Stratton also was impressed by Fiske. Such books by the philosopher as *The Destiny of Man*, *Through Nature to God*, and *The Idea of God* were among the favorite works on his library shelf.[50]

Since the personal libraries of the bonanza kings usually

contained the most popular or standard works of the period, it is likely that other mining leaders were at least vaguely familiar with such social Darwinists as Fiske, William Graham Sumner, and Herbert Spencer.[51] In fact, 368,755 copies of the works of Spencer, the English philosopher and pioneer sociologist who largely formulated social Darwinism, sold in the United States between 1860 and 1900. Tremendously popular in America, Spencer's teachings were attractive to some American business leaders, who could thus feel that they were the ultimate byproducts of the evolutionary process.[52] After reading Spencer, Andrew Carnegie seemed almost reborn: "I remembered that light came in as a flood and all was clear. Not only had I got rid of theology and the supernatural, but I had found the truth of evolution."[53] Sumner, Spencer's American disciple at Yale University, also found adherents within the business community. Rich executives were reassured by his argument that "millionaires are a product of natural selection. . . . It is because they are thus selected that wealth—both their own and that entrusted to them—aggregates under their hands. They may fairly be regarded as the naturally selected agents of society for certain work. They get high wages and live in luxury, but the bargain is a good one for society."[54]

The western mining leaders could attribute their success not only to rugged individualism and other personal character traits, but to Nature's law and even God's will. Sumner conveniently fused the individualistic values of the Horatio Alger stories with the determinism of social Darwinism when he asserted that "'the strong' and 'the weak' are terms which admit of no definition unless they are made equivalent to the industrious and the ideal, the frugal and the extravagant. . . . Capital is only formed by self-denial, and if the possession of it did not secure advantages and superiorities of a high order men would never submit to what is necessary to get it."[55] The popular contemporary theologian Henry Ward Beecher attempted to reconcile social Darwinism with religion when he contended that "God intended the great to be great and the little to be little."[56] The most influential Protestant minister of his era and an admirer of Spencer, Beecher led the pro-Darwin movement in the church. According to a leading intellectual historian, "there could be no surer sign that evolution, with its theological formulations, was widely accepted."[57] Other prominent churchmen similarly acknowledged the providential purpose of social Darwinism. By proclaiming that "Godliness is in league with riches," William Lawrence, the Episcopal Bishop of Massachusetts, sanctified the business leaders who triumphed during the competitive Darwinian struggle for success.[58] In short, it was the individual, natural, and divine destiny of American business executives to succeed and lead society.

Social Darwinism and the entire ideology of success in business were based largely on a popular academic doctrine of the Gilded Age—laissez-faire economics. According to this view, which originated in such writings of the classical economists as Adam Smith's *Wealth of Nations* (1776), the economy would grow and people would prosper if businessmen were left alone to pursue their interests without government regulation. Government was supposed to maintain law and order and protect life and property, but not interfere in the conduct of business or personal matters. Ironically, while in theory the business community endorsed the concept of noninterference, in practice they insisted that government protect and support them with tariffs, land grants, and other subsidies in clear violation of the laissez-faire doctrine.

Like others in big business, some mining leaders were ambivalent regarding the role of government in the economy. Samuel Hauser, whose multiple Montana interests included banking, mining, and land reclamation, favored actions by the federal government which served his interests, but in the typical laissez-faire tradition opposed laws that hampered his freedom to act. Therefore, he reacted strongly against the federal government's efforts to enforce land laws controlling the cutting of timber on the public domain. Timber was essential to mining operations. Like his fellow Montana mine developer Marcus Daly, Hauser demanded the right to freely exploit such resources. In 1885, he and Daly wired Joseph K. Toole, Montana's Territorial Delegate to Congress, the following message: "We understand there is a move to stop cutting timber on public lands. If successful it will stop all the principal mines in this territory and throw out of employment thousands of people. The companies we represent support directly and indirectly 10,000 people and we know this would be disastrous."[59] Nonetheless, the federal government filed protracted timber trespass suits against Hauser, Daly, and William A. Clark. The latter was charged with fraudulently filing on 11,000 acres of timberland, but the U.S. Supreme Court denied the allegation on technical grounds. The government sued the Daly estate for $2 million for timber depredations in the 1890s only to lose the case in the U.S. Supreme Court.[60] According to the leading student of the subject, the correspondence of Daly, Hauser, and others reveals a "genuine indignation towards a remote, unapproachable and invincibly arbitrary government" and confusion regarding the intent of timber laws.[61]

The issue of the taxation of mineral lands also prompted the mining men to endorse government restraint in regulating business affairs. Controversy over whether to tax the proceeds or the assessed valuation of mining properties delayed the ratification of a state constitution in Nevada from 1863 until 1864. The 1863 document, which would have created a tax on all mines whether productive or not, was overwhelmingly defeated by the voters. In 1864, the constitutional convention delegates worked out a suitable compromise by authorizing the state to tax only the proceeds of mining properties.[62] In Utah, the same controversial issue was raised at the state constitutional convention of 1895. As a delegate, Thomas Kearns, whose mining fortune came largely from the Silver King Mine in Park City, Utah, argued for a tax only on net proceeds, implying that the other method would be confiscatory: "Are you going to put a tax on that man and sell out what he has put his life to accomplish—a claim that he has patented?"[63] His fellow Utah miner, Jesse Knight, protested: "A mine is unlike most other classes of property, such as land or factories, whose owners enjoy every year the use of the entire property. To tax the unrealized and unknown values in a mine is to tax a man on property which is not in his possession and which, in fact, may not exist. Even suppsng [*sic*] it were possible to ascertain how much ore is located in a given mine, who can tell what it will be worth when it is taken out?"[64] Knight concluded that "no particular class of property could be taxed out of existence."[65]

Many Montana mine owners would have agreed with Knight. In the abortive state constitutional convention in 1884 and in the 1889 convention which resulted in statehood, the Montana delegates decided to exempt mining property from taxation, although the annual net proceeds of such property could be taxed by the state legislature. While there is little evidence that the mining interests of Montana, including the Clark and Daly forces, conspired in 1889 to obtain special taxation privileges, it is obvious that the decision of the constitutional convention was consistent with the laissez-faire doctrine against governmental meddling in the economy.[66]

At the same time, other mining elites took positions that directly violated the integrity of laissez-faire capitalism. In Colorado, where the rates charged by the railroads and water companies servicing irrigated farms became a politically volatile issue, Horace Tabor apparently thought government should control the inevitable tendency of companies to monopolize a given industry: "We have got too many of these eat-up Companies in America today and we do not want any more. There is entirely too much pooling in America."[67] Nathaniel P. Hill, the founding father of Colorado's smelting industry, claimed that competition "is a wholly inadequate protection against the excessive charges of railroad, telegraph, telephone, and similar companies."[68] He endorsed the U.S. Supreme Court ruling in *Munn v. Illinois* (1877) on behalf of the right of state government to regulate private property devoted to a public use, claiming that "to the extent that private property is used in such a way as to concern the whole public... it... ceases to be a matter of merely individual right."[69] His position was in part self-serving, as Hill himself indirectly admitted: "The owners of farms, of mines, and of factories are obliged to pay to monopoly in enhanced rates of transportation and telegraphing, and in other ways a tax upon their capital, of which only a small part can be shifted off upon the laborers whom they employ."[70] Jerome Chaffee, a fellow Colorado mining entrepreneur and former U.S. Senator from the state, was sufficiently concerned about the monopoly issue to virtually denounce his own Republican party in 1882 and then threaten to run again for public office. Believing the party to be controlled by the railroads, he said: "The time has nearly come when we must determine whether we are a nation of men or belong to the railroad companies. I have made up my mind that unless we get a white man for president next time I shall in my part of the world try the force of revolt. I shall run for governor as the anti-monopoly candidate, and I can carry Colorado."[71] Thus Tabor, Hill, and Chaffee—three of Colorado's more prominent mining entrepreneurs—did not accept the absolute inviolability and sanctity of private property. Laissez-faire economics notwithstanding, government had a responsibility to regulate powerful public utilities and some private corporations that had become monopolistic.

This notion gained greater support during the Progressive reform era after the turn of the century. In 1906, during his unsuccessful campaign to become Colorado's U.S. Senator, Thomas Walsh made a statement that seemed unusual for a millionaire businessman even during a reform period:

> Accumulated and concentrated wealth, both corporate and individual, is crushing from the masses the life of individual ownership, individual independence and, almost, individual existence. . . . A continuance of these monopolistic conditions means that all the commercial wealth of our country will be concentrated in a few hands, and the masses of the people left without an atom of ownership. We should establish a new basis for compensating the masses.[72]

He proposed a federal income tax, federal life insurance, and government savings banks as possible solutions to the problem. His daughter recalls that her father's political program was labelled "modified socialism" by the *Denver Post*.[73] If so, his suggestions were a striking departure from the economic philosophy of laissez-faire capitalism.

While some mining leaders called for the government to act against monopolies, others insisted that it tackle organized labor. The laissez-faire doctrine that government should at least protect property and maintain law and order was raised dramatically in labor-management disputes. Although the mining magnates on the whole were more responsive to the needs of their workers than is commonly assumed, they were not reluctant to call upon law enforcement to quell militant labor unrest when it violated property rights or threatened anarchy.[74]

John Hays Hammond, the internationally known mining engineer, advisor to such mine owners as George Hearst, Cecil Rhodes, and the Guggenheims, and a western mine owner in his own right, supported the right of workers to organize in order to protect themselves from organized capital, but he was quick to distinguish between acceptable and unacceptable unions:

> I believed in the Railway Brotherhoods and also in the American Federation of Labor under the guidance of. . .

> Samuel Gompers. I supported both organizations as valuable parts of our industrial system. On the other hand, I had no use for the radicalism of the IWW [Industrial Workers of the World], organized by the Western Federation of Miners (WFM) in 1905, of which I had had experience in. . . Coeur d'Alene [Idaho] and Cripple Creek [Colorado]. . . . If a government does not honestly and speedily enforce the laws made to protect both individuals and corporate bodies against attacks on life and property, it invites disaster upon itself.[75]

During the WFM strike at Cripple Creek in 1904, Hammond told a member of President Theodore Roosevelt's cabinet that the president should have complied with the Colorado governor's request to send troops to suppress the "lawless activities."[76]

Evidently, Hammond's mining-engineer colleagues were even more adamant than he in opposing working-class radicalism. They viewed the WFM, which by the turn of the century had broken with the American Federation of Labor and endorsed a Marxist-socialist ideology, as a "dictatorial and irresponsible body seeking to overthrow the capitalist system, and. . . compared the Butte Miners' Union—'a gang of Irish-Austrian-Italian anarchists'—to the dread Mafia."[77] During an eight-hour strike at his plant in 1903, James B. Grant, the German-trained mining engineer who served as vice president of the Omaha and Grant Smelting and Refining Company in Denver, charged the WFM with an insatiable desire to reduce the length of the working day: "We are in the fight and we'll be there at the finish. What is the use of giving in? The Western Federation of Miners now want eight hours. If we grant them that it will only be a question of time before they are striking for six and if it should get that it will be clamoring for four."[78]

Irving Howbert was especially critical of state governments that cooperated with the WFM rather than with the mine owners. In 1893–94, local unions in Cripple Creek became affiliated with the WFM and struck the mines owned by Howbert, David H. Moffat, and other leading Colorado entrepreneurs who refused to grant a three-dollar, eight-hour day. Cripple Creek soon became an armed camp. Moffat and his practical mining engineer-partner Eben Smith led a movement to collect funds to hire the notorious El Paso County sheriff's deputies to drive the miners from the properties they had seized. This strategy failed.[79] According to Howbert, "the striking miners had virtual possession of the camp. They refused to recognize the legally constituted authorities, and by their actions brought about a condition of anarchy that continued for a number of months."[80] The Populist governor of Colorado, Davis H. Waite, intervened with the militia and settled the conflict on terms favorable to the striking miners. In his memoirs, Howbert complained: "Governor Waite was a most erratic person, thoroughly imbued with socialistic and other fantastic ideas as to matters of government, and had unlimited confidence in his own opinions."[81]

In 1894, Howbert, the chairman of the Colorado State Republican Central Committee, succeeded in getting a Republican who would respect the rights of private property elected to the governor's office and in "redeeming the State from the radical Populistic government" which had advocated, among other reforms, the eight-hour day. In Colorado and other western mountain states, populism was in part an urban working-class movement which received support from the WFM.[82] The mining men as a group were not willing to endorse a political party that identified with the working class and called for a comprehensive reform program that would substantially expand the role of the federal government in the economy and threaten their own economic interests. In fact, according to Gene Clanton, "perhaps populism was, in the final analysis, America's aborted attempt at the creation of a truly radical and democratic working-class movement."[83] Edward Wolcott, Colorado's U.S. senator and the brother of mine owner Henry R. Wolcott, undoubtedly expressed the reservations of many bonanza kings when he wrote Howbert on 28 April 1896: "The Populist party . . . united with its free-coinage advocacy socialistic and paternalistic doctrines which are dangerous in tendency and which would be, if adopted, destructive of free institutions. . . . There is in my opinion one event which might involve our country in worse disaster than gold monometallism, and only one, and that would be the triumph of Populism."[84] Only one of the mining elites, Adolph Sutro, who is best remembered for constructing the Sutro Tunnel to allow better access and improve the ventilation and drainage for the Comstock mines, was closely associated with the party. He served as Populist reform mayor of San Francisco from 1895 until 1897, primarily in opposition to the Southern Pacific Railroad's influence in the city.[85] Although the mining elites were not sympathetic with populism, a number joined the free silver crusade the Populists championed. As silver miners or those who represented silver mining constituencies as elected public officials, they saw the free and unlimited coinage of silver as a legitimate governmental subsidy of business regardless of laissez-faire dogma.

At the official government ratio of sixteen to one, whereby sixteen ounces of silver equaled in value one ounce of gold, silver had been undervalued ever since the California gold rush of 1849 had lowered the price of gold. Thus, when Congress demonetized silver, it was more profitable for mine owners to sell the metal commercially on the open market than to the United States mint. Shortly thereafter, mines in Nevada, Arizona, Utah, and Colorado produced such large quantities of silver that the price fell on the world market, thereby making remonetization an attractive goal. However, the first drive to restore silver currency in the 1870s did not originate with the familiar coalition of mine owners and agrarian inflationists that was to characterize the free silver politics of the 1890s. According to Allen Weinstein's revisionist thesis, the early movement for remonetization was essentially an urban crusade with initial bases of support in the large towns and cities of the Midwest and Middle Atlantic states. A loosely formed combination of hard money newspaper editors, businessmen, academic reformers, bankers, and commercial groups comprised the first bimetallist leaders. Western legislators and prominent owners of silver mines, at least in the 1870s, remained indifferent to remonetization.[86] This was especially true of the Comstock kings—John Mackay, James Fair, James Flood, and William O'Brien—the largest producers of western silver. Under the terms of the Specie Resumption Act of 1875, for several years the U.S. Treasury eagerly purchased much of their silver, so they had little reason to support remonetization to ensure an adequate government market for their available supply of bullion.[87]

John P. Jones was a notable exception in the pattern of initial indifference to remonetization among leading western silver producers. A Comstock silver mine owner and Nevada's U.S. senator from 1873 to 1903, he promoted the issue more to advance his political career and out of ideological considerations than for reasons of economic self-interest.[88] According to Weinstein, he "played a pivotal role in the emergence of silver as a national issue in 1876, helping to propagate doctrines later identified with the entire mining West."[89] In the Senate, Jones served on the committee on mines and the important Finance Committee. He chaired the monetary commission established by joint congressional resolution on 15 August 1876, and wrote the commission's majority report, which analyzed the history of precious metals and the causes of the change in the relative value of gold and silver and recommended the restoration of bimetallism in the United States.[90] He was the Senate's most aggressive advocate of the Bland silver bill after the measure had passed the House in November, 1877.[91] One scholar asserts that Jones's "major speeches in behalf of the free coinage of silver—clear, informed, and reasonable—were accounted the best positions on the subject to be found anywhere."[92] Passed over the veto of President Rutherford B. Hayes in 1878, the Bland-Allison Act authorized the Treasury to purchase not less than $2 million and not more than $4 million worth of silver each month at the current market price and coin it into dollars. However, the act did not fully satisfy the free silverites because the Treasury always bought the minimum amount of silver, and its market price continued to decline until its ratio to gold was twenty to one by 1890 rather than the celebrated free-silver ratio of sixteen to one.[93]

The national depression of the early 1890s apparently reaffirmed Jones's faith in free silver as an economic panacea. While serving in the Senate in 1894, he wrote to his wife: "The working people of the country were never in such a deplorable condition as they find themselves in today and while now and again the dreadful strain may relax a little, I can see no chance for an era of real prosperity until a reform in the monetary policy of the country such as I have long contended for shall be inaugurated."[94] Another silver miner, Horace Tabor of Colorado, left this appraisal of Jones' efforts in the Senate in behalf of free silver and those of the other Nevada senator, the noted mining lawyer and political agent of the Southern Pacific Railroad William M. Stewart: "I agree with Jones in his measures. He is very sound. I am prepared to endorse all that Jones said; and Senator Stewart is also a very able minded man. . . . The best men we have got there [in the U.S. Senate] are now Jones and Stewart. Jones, Stewart and me harmonize on silver. There is only one difference. They get a chance to vote on it where I don't."[95]

Nonetheless, Tabor did not assume a passive role in the free-silver crusade. In 1885, he served as a delegate to the National Silver Convention held at his own Tabor Grand Opera House in Denver. In 1889, the Denver Chamber of Commerce hosted a state silver convention which resulted in the creation of the Colorado Silver Association with Tabor as president. As such, he urged President-elect Benjamin Harrison to appoint a free-silver advocate as Secretary of the Treasury. In 1890, the silverites secured passage of the Sherman Silver Purchase Act despite Harrison's opposition to remonetization. The act authorized the Treasury to purchase four and a half million ounces of silver a month and to issue notes redeemable in gold or silver coin. Although the legislation appeased some, others found it insufficient. No provision was made for unlimited coinage of the metal and the silver was purchased at the current low market price rather than at the former sixteen-to-one ratio. By 1892, the average value of an ounce of silver had dropped to $.87 from $1.32 in 1872. Of interest to inflationists, the Sherman Act produced no more currency expansion than had its predecessor, the Bland-Allison Act.[96]

As the Republican presidential nominating convention of 1892 approached, Senator Henry M. Teller of Colorado wrote Tabor: "If we are to support Mr. Harrison in convention as our first choice it is a declaration that we care but little about the free coinage of silver."[97] Apparently Tabor had not lost faith in the Republicans. In February 1892 he met with President Harrison and Secretary of State James Blaine to discuss the silver question, and overconfidently reported to his wife: "I am sure I have done good for the silver cause here. . . . My reception by them all was magnificent."[98]

Tabor was not the only Colorado miner-politician to support free silver. Nathaniel P. Hill, who represented Colorado in the Senate from 1879 until 1885, also championed the cause. His speech to the Senate on the issue for 20 June 1882 received favorable comment in the reputable *London Economist*. His 14 December 1884 address also received critical praise. Hill continued to work for the cause of free silver out of office as a member of the International Monetary Commission in 1891.[99]

Simeon G. Reed's silver mining interests in Idaho encouraged him to contribute his influence to the free-silver campaign. A letter to his business secretary from New York early in 1889 suggests that he tried to secure support for the remonetization of the metal from the U.S. Senators representing Oregon:

> As our Legislature meets the second Monday in Jany. . . I consider it very doubtful if I get home by that time. I presume Mr. [Joseph] Dolph will receive the nomination and will be re-elected Senator, but I understand he is opposed to the remonetization of Silver, and that he is the *only* Representative in Congress west of the Rocky Mountains, that is not strongly in favor of the Silver question—and it is very important to the Coast that when a question of great importance comes up that we should have a *united delegation* from the Pacific Coast. . . . The easiest way to get at this matter, (which is of so *vital* importance to the wheat producers of Oregon, as well as the producers of Silver) is to have the Legislature (before electing a Senator) instruct our Senators to vote for the remonetization of Silver.[100]

Reed found an ally on the silver issue in eastern banker Jay Cooke. As the financial agent for the Northern Pacific Railroad, Cooke had briefly held controlling interest in the Oregon Steam Navigation Company, formerly owned by Reed. He joined the latter in lobbying for free silver before members of Congress and the executive branch.[101] Reed went a step further by securing authorization from the secretary of state of Oregon on 12 November 1889 to serve as a delegate from Oregon to the National Silver Convention in St. Louis, which declared for the unrestricted coinage of silver.[102]

The National Silver Convention was the forerunner of another national pressure group, the American Bimetallic League, charged with the responsibility of securing federal free-coinage legislation. Strongly supported by the silver miners, this organization pursued a politically nonpartisan course. By 1892, it had gained a majority of votes for a free-silver bill in the Senate in part because the silver-mining West was disproportionately represented there. However, it had failed to win sufficient support in the House of Representatives. Its work was severely threatened by congressional repeal of the Sherman Silver Purchase Act, which President Grover Cleveland blamed for the financial panic in 1893. This fateful year also saw league-sponsored free-silver conventions in Washington, D.C., St. Louis, and Chicago. To the latter session came the Utah silver mining magnate Thomas Kearns, such national leaders as William Jennings Bryan, Terence Powderly, and Ignatius Donnelly, and one hundred members of the league-affiliated Montana Free Coinage Association, which Marcus Daly and Samuel Hauser served, respectively, as president and chair of the executive committee. In fact, Daly contributed so much money to the league that reportedly he was its chief financial supporter. Although the league continued to sponsor free silver conferences, lectures, and legislative petitions and bills before Congress, it was obvious by 1896 that the Democratic party offered the last best hope for a political victory necessary to stop the rapid decline in the price of silver since 1892.[103]

In 1896, the Republican party convention nominated William McKinley for president and adopted a solid gold platform. This resulted in the defection of thirty-four delegates from western silver-mining states. Led by Senator Teller, these silverites walked out of the convention and helped to organize the National Silver Republican Party. In a national convention at St. Louis in late July, the new party endorsed the Democratic presidential and vice-presidential nominees. Earlier in the month, the Democrats had nominated William Jennings Bryan for president on a platform calling for the free and unlimited coinage of silver at the sixteen-to-one ratio.[104] This action prompted some silver mining and smelting entrepreneurs, who had formerly affiliated with the Republicans, to cross party lines in their own economic self-interest.

In Colorado, Irving Howbert informed Teller that Bryan was "too much tainted with populism to suit me, but the issue is made and we must support him."[105] Although David H. Moffat considered himself a Republican, he voted for Bryan in 1896. However, in 1900, he gave $25,000 to defeat Bryan. In a telegram of 3 October 1900 to Colorado's Senator Edward O. Wolcott, he protested: "Certainly no good can come to us through Mr. Bryan's election..I voted for him four years ago and have regretted it ever since."[106] Nonetheless, Moffat's earlier 1896 electoral decision was shared by his fellow Colorado Republican, Hill. He deserted the party over the silver issue and joined his former political rival, Teller, in the new silver political organization.[107] On the other hand, Hill's smelting partner, Henry R. Wolcott, refused to abandon the Republican party in 1896. Judging by a speech he made while campaigning for governor of Colorado in 1898, he found free silver and Republicanism potentially compatible: "We need greater activity in our mines and in our works. . . . Business is to be improved in every direction, and this can be accomplished in my judgment, only through the Republican party. It is through that party alone that we can ever expect to see silver restored to the position which it must sooner or later again occupy as a money metal the world over."[108]

In Utah, Republican party activist Thomas Kearns contributed substantially to Bryan's 1896 campaign. He explained: "I am still a Utah Republican, but I will support no gold man."[109] By 1900, Kearns, like Moffat, had returned to the Republican party, serving as a delegate to the national convention in that year.[110]

In Nevada, the state Silver party, organized in 1892, enlisted the support of most of the leading local figures, including Republican senators Jones and Stewart. The former explained his political defection of 1894: "Both the old party organizations persist in treating this most imperious of all questions [free silver] as a mere tail to the party kites."[111] The new party also attracted many other Republicans and most Democrats. In 1896, the Silver and Democratic parties fused in support of Bryan, while retaining their separate political identities. In that year, Jones won re-election to the U.S. Senate with the endorsement of the Silver-Democrats. At the same time, he, like other silver miner–politicians, campaigned for Bryan.[112] His admiration for the "Great Commoner" endured beyond the free-silver campaign of 1896. In 1899, he wrote to his wife that Bryan "is a wonderfully clever man. The press will have to give up the attempt to ridicule him. He has proved himself equal to every occasion."[113]

In Montana, prominent Democratic capitalists who owned silver-producing mines eagerly joined the movement to nominate and elect Bryan in 1896. Hauser, who as territorial governor of Montana in 1885 had written a report to the U.S. secretary of the interior calling for free silver, led the Montana delegates to the Democratic nominating convention. The delegation happily backed Bryan's candidacy.[114] After the nomination, such other Montana mining magnates as Daly, Clark, and perhaps Augustus F. Heinze contributed to Bryan's campaign. Although some have suggested that Daly donated as much as $300,000 to the 1896 campaign, precise dollar amounts are unknown. The most recent student of the Montana copper kings suggests that "Daly . . . put forth a great deal of money. Combined with lesser amounts from Clark and others, it made Montana a key source of Democratic-Populist funding."[115]

Evidently, the vast majority of the bonanza kings with an economic or political stake in silver mining supported Bryan in 1896. In some states, such as Montana and Idaho, this meant that they backed a fusion ticket consisting of Populist and Democratic candidates. Having sacrificed their political integrity and broad reform program for the sake of free silver, the Populists nominated Bryan for president and fused with the Democrats in

1896.[116] It should not be assumed, however, that all mining entrepreneurs simply voted their economic self-interest by supporting Bryan in 1896. Winfield S. Stratton of Colorado not only came out for Bryan, but he offered to back up his choice with a $100,000 bet if someone would put up $300,000 on McKinley. As a prominent Cripple Creek gold mine owner, he had nothing to gain and much to lose from a free-silver victory. If he won the bet, the $300,000 would be donated to the Colorado Springs Free Reading Room and Library Association. No one accepted his apparently sincere offer. As Stratton explained: "I do not make the offer because of any information that I have on the election, but I have a feeling that Bryan is going to win. I am deeply interested in seeing Bryan elected. I realize that the maintenance of the gold standard would perhaps be best for me individually, but I believe that free silver is the best thing for the working masses of the country. It is because I have a great respect for the intelligence and patriotism of the working people and I believe that they will see their duty at the polls that I am willing to make such an offer."[117] Stratton was atypical of the mining elites. In an age when many leading capitalists believed that the political process was designed primarily to serve private business, such an action in behalf of the working class and against one's own interest was highly peculiar.

After Bryan's electoral defeat in 1896, the free silver issue was no longer politically viable. Huge new gold discoveries in Canada, Australia, South Africa, and Alaska inflated the currency supply and helped raise prices, much to the delight of those inflation-minded farmers who had supported the Populist movement. The renewed prosperity restored widespread faith in the gold standard. Silver mining entrepreneurs such as Moffat of Colorado, Kearns of Utah, and Jones of Nevada, who abandoned the Republic party to back Bryan, had resumed their former political ties by the beginning of the twentieth century. In fact, in October 1901, Jones informed a reporter for the *Nevada State Journal* that he would support a Republican candidate for U.S. senator "with all my heart." He had decided that the silver issue was "dead."[118] Despite the ultimate failure of the free-silver movement, a number of western mining elites had tried various political means to convince the federal government to restore the free and unlimited coinage of silver at the former sixteen-to-one ratio with gold.

Since lead was often found in combination with silver, many mine owners supported a tariff on the importation of competitive foreign lead. Moffat undoubtedly shared the sentiments of one of his chief employees on this subject. In late 1884, the manager of his Henriett Mine wrote George Kassler, Moffat's business associate: "Am afraid the free trade tendencies of the Dimmycrats [*sic*] will hurt our lead business."[119] Although the United States lead industry had traditionally been protected by tariffs, Moffat and other lead producers had reason to be concerned.

In 1888, the Treasury Department during the Democratic Cleveland administration decided that lead ore would be defined as ore in which lead values exceeded those of any other metal. This meant that large amounts of lead ore in which silver was the predominant value could be imported from Mexico without a duty assessed on the lead content. The cheaper Mexican lead created a temporary surplus and threatened the operation of lead-silver mines in Idaho, Utah, and Colorado.[120] In 1888, Simeon G. Reed, who had a large stake in lead production as owner of the lead-silver rich Bunker Hill and Sullivan, complained to U.S. Senator Stewart of Nevada: "This shipping of Lead ores containing Silver into the Country under the ruling of the Secty of the Treasury is working a great hardship on our Mining interest in the Northwest, and I am pleased to learn the matter is being taken hold of in Congress by one so well posted in such matters."[121] In another communication with Stewart, Reed offered to come to Washington, D.C., to lobby for a change in policy.[122]

When Congress considered the entire issue of protection for domestic lead, Reed sought to influence Oregon's senator, Joseph Dolph:

> Should the tariff be reduced it would close down all the low grade Lead producing Mines in the Country, and only the high grade Ores could be worked at a profit—New York is the market for Lead, and even at the minimum rate of freight charged by the Railroads the distance from the Mining Centers is so great in connection with the Smelting and reduction charges that a small reduction in the tariff would greatly effect the output, and open the market to foreign Lead. . . . In fact there can hardly be (as it seems to me) any argument that can be used in favor of a reduction of the tariff on Lead, and still these "Cleveland Democrats" may try it on.[123]

In 1889, Reed received a request from the Republican State Committee of Washington, urging him to exercise his influence to secure the appointment of William McKinley as Speaker of the House of Representatives in order to protect western lead and other interests.[124] Although Thomas B. Reed (who was no relation to Simeon Reed) won the speaker's chair, McKinley was quite important to western mining interests. As chairman of the House Ways and Means Committee from 1889 to 1891, he was largely responsible for drafting the tariff act of 1890 (the McKinley Tariff), which raised duties on a number of imported items to the highest levels up to that time. The tariff included a duty of $0.015 per pound on all imported ores containing lead, effectively shutting out competition from Mexico.[125]

Later, when the tariff on lead again became a matter for political debate, Thomas Kearns and Jesse Knight of Utah made their interests known as important lead-silver producers. When the price of silver declined during the depression of the early

1890s, lead protectionism became more critical than ever before. In 1893, following a June layoff of workers at his Silver King Mine, Kearns asserted that if the tariff were to be taken from lead it would "knock them [the mines] out in short order," with the silver market depressed.[126] As a delegate from Summit County to the Utah State Republican Convention in 1895, he strongly supported maintaining the tariff.[127] In 1909, when the mining industry again faced a possible reduction in the lead tariff, Knight telegraphed Senator Nelson Aldrich of Rhode Island that "any further reduction in [the] tariff on lead would necessitate the closing down of many mines and smelters and cause general panic in the West."[128] As chairman of the powerful Senate Finance Committee, Aldrich was responsible for numerous amendments to a House-sponsored tariff bill which proposed substantial rate reductions on many items. The controversial Payne-Aldrich Tariff Act (1909) actually raised rates and removed items from the free list.[129]

The political behavior of the western mining elites suggests that one did not succeed in business simply by having the right kind of character or by having been born relatively poor. In fact, outstanding achievement in the mining industry required an ability to effectively and often unscrupulously manage such factors of production as property, capital, labor, technology, and government.[130] Nonetheless, the mining leaders largely subscribed to the popular simplistic ideology of the period which suggested that young men who applied the Horatio Alger–style virtues of hard work, thrift, and sobriety would achieve success. The relatively modest social origins of the vast majority of the mining magnates seemed to confirm this formula for vertical social mobility if not the actual leap from rags to riches. In addition, it is likely that some of the mining elites rationalized their success as the inevitable working of nature's law, given the acceptance by some leading American businessmen of social Darwinism. Also, these wealthy rugged individualists could take further comfort from the argument of religious leaders that God was presumably on their side.

Their attitudes about the proper role of government in the economy were ambivalent. They honored the theory of laissez-faire capitalism by urging government restraint in the taxation of mineral lands and in the enforcement of timber-cutting regulations. However, they insisted upon government regulation when it served their interests, such as in the policing of corporate monopolies or public utilities such as railroads. At the same time, they demanded that government create a favorable climate in which to conduct business by suppressing labor unrest and instituting the free and unlimited coinage of silver and a protective tariff on the importation of foreign lead. Paradoxically, the cult of the self-made man, social Darwinism, and government subsidy formed the ideological core of business success as perceived and practiced by the mining entrepreneurs. In this regard they were similar to leaders of other industries who ostensibly endorsed rugged individualism, natural selection, and laissez-faire capitalism, but looked when necessary to government for support and protection. While these so-called robber barons had mixed attitudes about the role of government in the economy, they were more consistent in financially aiding such governmental functions as education.

NOTES

1. Wilbur Fisk Stone, ed., *History of Colorado*, vol. 1 (Chicago: S. J. Clarke Co., 1918), 296. Walsh's career can be traced in Evalyn Walsh McLean, *Father Struck It Rich* (Boston: Little, Brown and Co., 1936), and in Walsh's business correspondence (1887–1903), Evalyn Walsh McLean Family Papers, 1886–1948, Manuscript Division, Library of Congress, Washington, D.C. (hereafter cited as Walsh Papers).
2. *Denver Post*, 10 April 1910.
3. McLean, *Father Struck It Rich*, 36.
4. *Rocky Mountain News* (Denver), 2 April 1910.
5. Ibid.

6. McLean, *Father Struck It Rich*, 93.

7. Knight to Victor T. Petterson, 4 December 1908, box 50, 1908–9, file P, Business Records and Family Papers of Jesse Knight, Brigham Young University Archives, Provo, Utah (hereafter cited as Knight Papers). Knight's career is profiled in Gary Fuller Reese, "'Uncle Jesse,' The Story of Jesse Knight, Miner, Industrialist, Philanthropist" (M.S. thesis, Brigham Young University, 1961).

8. Horace A. W. Tabor, "Life of Horace Tabor" (autobiography), 1889, Bancroft Library, University of California, Berkeley, 2, 5. Tabor's controversial career is covered in Duane A. Smith, *Horace Tabor: His Life and the Legend* (Boulder: Colorado Associated University Press, 1973).

9. As quoted in G. W. Barrett, "When Big Money Came to Owyhee," *Idaho Yesterdays* 13 (Spring 1969): 25.

10. Quoted from the *Denver Polly Pry*, ca. 1914, newspaper clipping in Thomas F. Dawson, comp., Scrapbooks, State Historical Society of Colorado, Denver, 59:289.

11. Irvin G. Wyllie, *The Self-Made Man in America: The Myth of Rags to Riches* (New York: The Free Press, 1966), 40.

12. As quoted in Cecil G. Tilton, *William Chapman Ralston, Courageous Builder* (Boston: Christopher Publishing House, 1935), 308–9. Of the several accounts of Ralston's career, the best is David Lavender, *Nothing Seemed Impossible: William C. Ralston and Early San Francisco* (Palo Alto: American West Publishing Co., 1975).

13. Simeon Wenban, Dictations and Biographical Material, 1888, Bancroft Library, University of California, Berkeley, 3.

14. Hubert H. Bancroft, *Chronicles of the Builders of the Commonwealth*, vol. 4 (San Francisco: History Co., 1891-92), 246.

15. Dennis Sheedy, *The Autobiography of Dennis Sheedy* (Denver: privately printed, 1922), unnumbered pages in the Foreword.

16. Ibid., 7.

17. Ibid., 8.

18. As quoted in William Randolph Hearst, Jr., "The Frontier Life—Part 2," *San Francisco Examiner and Chronicle*, 21 August 1966.

19. Ibid.

20. As quoted in William Randolph Hearst, Jr., "Thinking of Pop After 15 Years," *San Francisco Examiner and Chronicle*, 14 August 1966.

21. As quoted in Edwin Wildman, *Famous Leaders of Industry: The Life Stories of Boys Who Have Succeeded* (Boston: The Page Co., 1920), 76.

22. For the Clark-Daly feud, see Carl B. Glasscock, *The War of the*

Copper Kings, the Builders of Butte and the Wolves of Wall Street (New York: Grosset and Dunlap, 1966 [orig. pub. 1933); and Michael P. Malone, *The Battle for Butte: Mining and Politics on the Northern Frontier, 1864–1906* (Seattle: University of Washington Press, 1981), especially 80–110.

23. Michael P. Malone, "Midas of the West: The Incredible Career of William Andrews Clark," *Montana: The Magazine of Western History* 33 (Autumn 1983): 4.

24. As quoted in Charles C. Goodwin, *As I Remember Them* (Salt Lake City: Salt Lake Commercial Club, 1913), 122.

25. For the protests of unemployed workers, see Carlos A. Schwantes, *Coxey's Army: An American Odyssey* (Lincoln: University of Nebraska Press, 1985).

26. Goodwin, *As I Remember Them*, 330. For Packard's career, see J. M. Guinn, *History of the State of California and Biographical Record of the Sacramento Valley, California* (Chicago: Chapman Publishing Co., 1906 [orig. pub. 1902), 360–63; and *National Cyclopedia of American Biography*, vol. 16 (New York: James T. White and Co., 1892-1969), 411.

27. Carroll Van West, "Montana's Self-Confessed Elite: The Progressive Men of Montana," *Montana: The Magazine of Western History* 33 (Autumn 1983): 36.

28. Harvey W. Scott, ed., *History of Portland, Oregon, with Illustrations and Biographical Sketches of Prominent Citizens and Pioneers* (Syracuse: D. Mason and Co., 1890), 527. For Reed's mining career, see Richard H. Peterson, "Simeon Gannett Reed and the Bunker Hill and Sullivan: The Frustrations of a Mining Investor," *Idaho Yesterdays* 23 (Fall 1979): 2–8.

29. William N. Byers, *Encyclopedia of Biography of Colorado* (Chicago: Century Publishing and Engraving Co., 1901), 247. For Stratton's career, see Frank J. Waters, *Midas of the Rockies: The Story of Stratton and Cripple Creek* (Denver: Alan Swallow, 1949 [orig. pub. 1937).

30. Byers, *Encyclopedia of Biography of Colorado*, 207.

31. Ben A. Block, comp., *Colorado—Its Resources—Its Men* (Denver: State of Colorado, 1901), 45.

32. Henry R. Wolcott, Dictation, 1888, Bancroft Library, University of California, Berkeley, 3.

33. As quoted in Thomas F. Dawson, *Life and Character of Edward Oliver Wolcott, Late a Senator of the United States from the State of Colorado*, vol. 1 (New York: Knickerbocker Press, 1911), 80.

34. As quoted in *The Financier* (New York), 17 October 1885, in

Henry R. Wolcott, Dictation and Related Biographical Material for David H. Moffat, 1885, Bancroft Library, University of California, Berkeley, 4.

35. Steven F. Mehls, "David H. Moffat, Jr.: Early Colorado Business Leader" (Ph.D. diss., University of Colorado, Boulder, 1982), 2.

36. *The Progressive Men of the State of Montana* (Chicago: A. W. Bowen and Co., 1901), 18.

37. George Hearst, *The Way It Was: Recollections of U.S. Senator George Hearst, 1820–1891* (San Francisco: The Hearst Corporation, 1972), 38.

38. See William Miller, "American Historians and the Business Elite," *Journal of Economic History* 9 (November 1949): 184–208.

39. Alonzo Phelps, *Contemporary Biography of California's Representative Men*, vol. 1 (San Francisco: A. L. Bancroft and Co., 1881), 325–28.

40. This explanation of how class membership was determined appeared originally in Richard H. Peterson, "The Frontier Thesis and Social Mobility on the Mining Frontier," *Pacific Historical Review* 44 (February 1975): 64. See pp. 61–63 and 66–67 for other relevant tables on the social origins of the mining men.

41. Dwight Akers, "David H. Moffat and His Home Town," *Colorado Magazine* 27 (July 1950): 213–14; Harold A. Boner, *The Giant's Ladder: David H. Moffat and His Railroad* (Milwaukee: Kalmbach Publishing Co., 1962), 11–12; and Edgar C. McMechen, *The Moffat Tunnel of Colorado, An Epic of Empire*, vol. 1 (Denver: Wahlgreen Publishing Co., 1927), 44. For a different assessment of Moffat's class standing, see Mehls, "David H. Moffat, Jr.," 1–3.

42. See especially Jonathan Bliss, *Merchants and Miners in Utah: The Walker Brothers and Their Bank* (Salt Lake City: Western Epics, 1983), 48–55. See also *Biographical Record of Salt Lake City and Vicinity* (Chicago: National Historical Record Co., 1902), 35–36; Orson F. Whitney, *History of Utah*, vol. 4 (Salt Lake City: George Q. Cannon and Sons, 1892–1904), 626–27; and Henry Hall, ed., *America's Successful Men of Affairs: An Encyclopedia of Contemporaneous Biography*, vol. 2 (New York: New York Tribune, 1895–96), 814–16.

43. Hall, ed., *America's Successful Men of Affairs*, 175.

44. As quoted in Degler et al., *The Democratic Experience*, 2:99.

45. Hearst, *The Way It Was*, 38.

46. See Richard Hofstadter, *Social Darwinism in American Thought*, rev. ed. (Boston: Beacon Press, 1955), 57.

47. Hubert H. Bancroft, "Biographical Sketch," in Dictation and

Biographical Material on Simeon G. Reed, 1888, Bancroft Library, University of California, Berkeley, 26.

48. Reed to Martin Winch, 23 December 1891, *Letters and Private Papers of Simeon G. Reed*, vol. 33 (Portland, Ore.: Reed College Project, Works Projects Administration, 1940), 239. (The originals of these letters are preserved by Reed College, Portland, and are hereafter cited as Reed papers.)

49. Degler et al., *The Democratic Experience*, 3d ed. (1973), 375.

50. Waters, *Midas of the Rockies*, 188.

51. For the provocative revisionist view that such Gilded Age conservatives as Spencer, Sumner, and Carnegie were not truly social Darwinists but Lamarckian evolutionists and that reformers and liberal academicians created a "myth" of conservative social Darwinism for their own ideological and political purposes, see Robert C. Bannister, *Social Darwinism: Science and Myth in Anglo- American Social Thought* (Philadelphia: Temple University Press, 1979).

52. See Thomas C. Cochran and William Miller, *The Age of Enterprise: A Social History of Industrial America* (New York: Harper and Row, 1961), 124–28; and John K. Galbraith, *The Age of Uncertainty* (Boston: Houghton Mifflin Co., 1977), 44–48. For the dissenting view that big businessmen of the Gilded Age were not necessarily conscious, philosophical social Darwinists, see Irvin G. Wyllie, "Social Darwinism and the Businessman," *Proceedings of the American Philosophical Society* 103, no. 5 (1959): 629–35.

53. As quoted in Degler et al., *The Democratic Experience*, 5th ed. (1981), 54.

54. As quoted in Hofstadter, *Social Darwinism in American Thought*, 58.

55. Ibid., 56–57.

56. As quoted in Galbraith, *The Age of Uncertainty*, 56.

57. Henry F. May, *Protestant Churches and Industrial America* (New York: Octagon Books, Inc., 1963 [orig. pub. 1949), 142.

58. William Lawrence, "Relation of Wealth to Morals," *World's Work* 1 (1901): 287.

59. As quoted in K. Ross Toole and Edward Butcher, "Timber Depredations on the Montana Public Domain, 1885–1918," *Journal of the West* 7 (July 1968): 354. See also John W. Hakola, "Samuel T. Hauser and the Economic Development of Montana: A Case Study in Nineteenth-Century Frontier Capitalism" (Ph.D. diss., Indiana University, 1961), 279–82.

60. Toole and Butcher, "Timber Depredations on the Montana Public Domain," 354–55; and Malone, *The Battle for Butte*, 43–45.

61. Toole and Butcher, "Timber Depredations on the Montana Public Domain," 360.

62. See Gordon M. Bakken, "The Taxation of Mineral Wealth and the Nevada Constitutional Convention of 1864," *Nevada Historical Society Quarterly* 12 (Winter 1969): 5–15; and Myron Angel, *History of Nevada with Illustrations and Biographical Sketches of Its Prominent Men and Pioneers* (Berkeley: Howell-North, 1958 [orig. pub. 1881), 28, 122–24.

63. As quoted in Kent S. Larsen, "The Life of Thomas Kearns" (M.A. thesis, University of Utah, 1964), 34.

64. *Deseret Evening News* (Salt Lake City), 12 October 1918.

65. Ibid.

66. See John W. Smurr, "The Montana 'Tax Conspiracy' of 1889," *Montana: The Magazine of Western History* 5 (Spring 1955): 46–53, and 6 (Summer 1955): 47–57.

67. Tabor, "Life of Horace Tabor," 21. For reactions against Colorado's railroads and privately owned water companies, see James Edward Wright, *The Politics of Populism: Dissent in Colorado* (New Haven: Yale University Press, 1974), 34–35, 39, 96–97.

68. Nathaniel P. Hill, "Address delivered at Leadville, Colorado, before the YMCA, April 12, 1888," in *Speeches and Papers on the Silver, Postal Telegraph, and Other Economic Questions* (Colorado Springs: Gazette Printing Co., 1890), 284. A former chemistry professor at Brown University, Hill is largely credited with the discovery in the 1860s that smelting—rather than the amalgamation process then in use—was required to extract gold from Colorado's sulfide or "refractory" ores. He put his idea to work successfully as the organizer and with Henry R. Wolcott, the manager of the Boston and Colorado Smelting Company. His career is examined effectively in James E. Fell, Jr., "Nathaniel P. Hill, a Scientist-Entrepreneur in Colorado," *Arizona and the West* 15 (Winter 173): 315–32. See also Hill's correspondence, Nathaniel P. Hill Papers, State Historical Society of Colorado, Denver, Colorado (hereafter cited as Hill Papers).

69. Hill, "Address delivered at Leadville, Colorado," 268.

70. Ibid., 287.

71. As quoted from unspecified newspaper interview, Dawson, Scrapbooks, Public Men, I, 61:5.

72. McLean, *Father Struck It Rich*, 133.

73. Ibid.

74. See Richard H. Peterson, "Conflict and Consensus: Labor Relations in Western Mining," *Journal of the West* 12 (January 1973): 1–17.

75. John Hays Hammond, *The Autobiography of John Hays Hammond*, vol. 2 (New York: Farrar and Rinehart, 1935), 699. Hammond apparently equated the IWW with the Western Federation of Miners (WFM).

76. Ibid., 52.

77. Clark C. Spence, "The Mining Engineers in the West," in *The American West: An Appraisal*, ed. Robert G. Ferris (Santa Fe: Museum of New Mexico Press, 1963), 109.

78. As quoted in Vernon H. Jensen, *Heritage of Conflict: Labor Relations in the Non-Ferrous Metals Industry up to 1930* (Ithaca: Cornell University Press, 1950), 111–12. For Grant's reaction to a similar strike in 1899, see the *Colorado Springs Gazette*, 11, 12 July; 16 August 1899.

79. Mehls, "David H. Moffat, Jr.," 137–38.

80. Irving Howbert, *Memories of a Lifetime in the Pikes Peak Region* (New York: G. P. Putnam's Sons, 1925), 271. For a description of the Cripple Creek strike, see Marshall Sprague, *Money Mountain: The Story of Cripple Creek Gold* (Boston: Little, Brown and Co., 1953), 133–56.

81. Howbert, *Memories of a Lifetime*, 272. For Waite's political career, see Marjorie Hornbein, "Davis Waite, Silver, and Populism," in *Essays and Monographs in Colorado History*, no. 1, ed. David N. Wetzel (Denver: Colorado State Historical Society, 1983), 71–90; and John R. Morris, *Davis H. Waite: The Ideology of a Western Populist* (Lanham, Md.: University Press of America, 1982).

82. Howbert, *Memories of a Lifetime*, 273–77. See also Melvyn Dubofsky, "The Origins of Western Working Class Radicalism, 1890–1905," in *Labor History* 7 (Spring 1966): 140–42. For evidence that populism in the Mountain West, at least until 1896, was a complex, multi-issue movement rather than one preoccupied simply with free silver, see Robert W. Larson, "Populism in the Mountain West: A Mainstream Movement," *Western Historical Quarterly* 13 (April 1982): 143–64; and Wright, *The Politics of Populism*, 126–82.

83. Gene Clanton, "'Hayseed Socialism' on the Hill: Congressional Populism, 1891–1895," *Western Historical Quarterly* 15 (April 1984): 161.

84. As quoted in Dawson, *Life and Character of Edward Oliver Wolcott*, 1:236.

85. Robert E. Stewart, Jr., and Mary F. Stewart, *Adolph Sutro; A Biography* (Berkeley: Howell-North, 1962), 202–9; and Archibald C. Unsworth, "Memoir" (material on Adolph Sutro), Sutro Branch, California State Library, San Francisco, 2:96–II, 148.

86. Allen Weinstein, *Prelude to Populism: Origins of the Silver Issue, 1867–1878* (New Haven: Yale University Press, 1970), 356.

87. Ibid., 150–76. "The Specie Resumption Act of January 15, 1875, contained a provision which called for the redemption of fractional greenbacks—ten, twenty-five, and fifty cent paper tokens—with subsidiary silver coins to be minted and issued 'as rapidly as possible' at the discretion of the Secretary of the Treasury" (ibid., 157).

88. Ibid., 53, 75–81.

89. Ibid. 53.

90. See U.S. Congress. Senate. U.S. Monetary Commission, *Report and Accompanying Documents*, Senate Report 703, 2 vols., 44th Cong., 2d sess. (Washington, D.C.: Government Printing Office, 1877–79).

91. Weinstein, *Prelude to Populism*, 70; and Mary E. Glass, *Silver and Politics in Nevada: 1892–1902* (Reno: University of Nevada Press, 1969), 17.

92. Gilman M. Ostrander, *Nevada, the Great Rotten Borough, 1859–1964* (New York: Alfred A. Knopf, 1966), 104. See also Harry M. Gorham, *My Memories of the Comstock* (Los Angeles: Suttonhouse Publishers, 1939), 111.

93. For a careful analysis of the evolution of the Bland-Allison Act, see Weinstein, *Prelude to Populism*, 301–53. Also on the act and free-silver issue, see Gene Clanton, "Silver Issue," in *The Reader's Encyclopedia of the American West*, ed. Howard R. Lamar (New York: Thomas Y. Crowell Co., 1977), 1110–12; and Richard B. Morris, ed., *Encyclopedia of American History* (New York: Harper and Row, 1970), 254, 261–62.

94. Jones to his wife, Georgina, 29 May 1894, box 3, folder 2, John P. Jones Family Papers, Special Collections Dept., University of California, Los Angeles (hereafter cited as Jones Papers).

95. Tabor, "Life of Horace Tabor," 17. For Stewart's career, including his work in behalf of free silver, see Russell R. Elliott, *Servant of Power: A Political Biography of Senator William M. Stewart* (Reno: University of Nevada Press, 1983). A laudatory description of Stewart's political activities is available in Effie Mona Mack, "William Morris Stewart, 1827–1909," *Nevada Historical Society Quarterly* 7 (1964): 5–121.

96. Smith, *Horace Tabor*, 304–5; Glass, *Silver and Politics in Nevada*, 20–21; Bancroft, *Chronicles of the Builders of the Commonwealth*, 4:329; Wright, *The Politics of Populism*, 134–35; and Morris, *Encyclopedia of American History*, 261–62.

97. Teller to Tabor, 15 February 1892, folio file 79, no. 7, Horace A. W. Tabor Family Papers, State Historical Society of Colorado (hereafter cited as Tabor Papers).

98. Tabor to Baby Doe (Elizabeth McCourt Doe Tabor), 17 February 1892, folio file 79, no. 8, ibid. See also Smith, *Horace Tabor*, 305.

99. Allen Johnson et al., eds., *Dictionary of American Biography*, vol. 9 (New York: Charles Scribner's Sons, 1928–58), 43; and Bancroft, *Chronicles of the Builders of the Commonwealth*, 4: 390–93.

100. Reed to Martin Winch, 4 January 1889, Reed Papers, 28:4–5. Reed was referring to Joseph N. Dolph, a former railroad lawyer, who served as U.S. Senator from Oregon, 1883–95. For information on the silver issue in Idaho, see Claudius O. Johnson, "The Story of Silver Politics in Idaho, 1892–1902," *Pacific Northwest Quarterly* 33 (July 1942): 283–96.

101. Cooke to Reed, 10 January 1889, Reed Papers, 28:11–12. For Cooke's relationship with the Oregon Steam Navigation Company, see Frances Fuller Victor, "Biographical Sketch of John D. Ainsworth," in Dictation and Biographical Material on Ainsworth, 1890, Bancroft Library, 24. Ainsworth was the president of the Oregon Steam Navigation Company.

102. Reed Papers, 29:90.

103. John D. Hicks, *The Populist Revolt: A History of the Farmers' Alliance and People's Party* (Lincoln: University of Nebraska Press, 1961 [orig. pub. 1931), 311–14, 342–43; Larsen, "The Life of Thomas Kearns," 30; and Malone, *The Battle for Butte*, 106.

104. Glass, *Silver and Politics in Nevada*, 105–9; and Morris, *Encyclopedia of American History*, 264–65.

105. As quoted in Wright, *The Politics of Populism*, 212 n. 79.

106. See Dawson, Scrapbooks, Business Men, I, 59:45; and the *Denver Republican*, 5 October 1900.

107. Dawson, *Life and Character of Edward Oliver Wolcott*, 1:175.

108. As quoted in ibid., 1:264. See also Dawson, Scrapbooks, Public Men, I, 61:47.

109. As quoted in Larsen, "The Life of Thomas Kearns," 42.

110. Ibid., 45.

111. As quoted in David J. Rothman, *Politics and Power: The United States Senate, 1869–1901* (Cambridge, Mass.: Harvard University Press, 1966), 70.

112. Glass, *Silver and Politics in Nevada*, 41–118; Ostrander, *Nevada, the Great Rotten Borough*, 114–19; and Samuel P. Davis, ed., *The History of Nevada*, vol. 1 (Reno: Elms Publishing Co., 1913), 426–28.

113. John P. Jones to Georgina Jones, 7 October 1899, box 3, folder 2, Jones Papers.

114. Malone, *The Battle for Butte*, 107; and Johnson et al., eds., *Dictionary of American Biography*, 8:403.

115. Malone, *The Battle for Butte*, 108–9. For a useful analysis of free-silver politics in Montana, see Thomas A. Clinch, *Urban Populism and Free Silver in Montana* (Missoula: University of Montana Press, 1970).

116. Hicks, *The Populist Revolt*, 370; and Malone, *The Battle for Butte*, 106–8.

117. As quoted in Sprague, *Money Mountain*, 210–11.

118. As quoted in Glass, *Silver and Politics in Nevada*, 201.

119. Harker to Kassler, 17 November 1884, in Agnes W. Spring, comp., "Notes and Extracts from the Personal Papers of David H. Moffat and Jerome B. Chaffee," State Historical Society of Colorado.

120. See Reed Papers, 23:355, 358–59; and "The Ruling of Lead," *Salt Lake Tribune*, 7 July 1888.

121. Reed to Stewart, 4 June 1888, Reed Papers, 23:198.

122. Reed to Stewart, July 1888, ibid., 23:205.

123. Reed to Dolph, 13 February 1888, ibid., 23:173–74.

124. John F. Gowey to Reed, 28 November 1889, ibid., 29:94.

125. The effect of the McKinley Tariff on imported lead is discussed briefly in John Fahey, *The Days of the Hercules* (Moscow: University Press of Idaho, 1978), 86. Reed was a regular contributor to the American Protective Tariff League. See Reed Papers, 33:209. For evidence that Sheedy of Colorado also worked to prevent the lead tariff from being reduced from $0.015 to $0.005 per pound, see Jerome C. Smiley, *History of Denver, with Outlines of the Earlier History of the Rocky Mountain Country* (Denver: J. H. Williamson and Co., 1903 [orig. pub. 1901), 564.

126. As quoted in Larsen, "The Life of Thomas Kearns," 18.

127. Ibid., 36.

128. Knight to Aldrich, 29 March 1909, Knight Investment Co. Correspondence, 1908–9, box 48, file A-B, Knight Papers.

129. For a brief description of the legislation, see *The Encyclopedia of American History* (Guilford, Conn.: The Dushkin Publishing Group, Inc., 1973), 255–56.

130. ee Richard H. Peterson, *The Bonanza Kings: The Social Origins and Business Behavior of Western Mining Entrepreneurs, 1870–1900* (Lincoln: University of Nebraska Press, 1977), 23–114.

2

The Spirit of Giving: Educational Philanthropy

THE AMERICAN BUSINESS LEADERS of the Gilded Age typically are identified as the "robber barons." Both in their own day and since, they have been charged with various kinds of industrial piracy, including issuing watered stock, bribing politicians, oppressing labor, fixing prices, and competing ruthlessly with their business rivals. Indeed, historians have compiled much evidence to support these indictments. On the other hand, some scholars have viewed the business elites of the period in a more positive manner, emphasizing their contributions to increased production, technological innovation, efficient organization, and the general material well-being of society. Such historians have seen the great capitalists of the last century as industrial pioneers and statesmen rather than as industrial profiteers. Their defense of the business leaders has been bolstered by the fact that some became well-known philanthropists in later life, thereby fulfilling important welfare and educational roles.[1] Without attempting to resolve the so-called "robber baron" controversy, a discussion of the educational philanthropy of prominent western mining entrepreneurs of the era should further our understanding of the behavior of business leaders and of western lifestyles. An analysis of charitable activities regarding colleges, universities, schools, libraries, and the arts will make possible an evaluation of the popular view that the business tycoons of the mining West were primarily interested in conspicuous consumption and were socially irresponsible to the point of having, as one historian asserts, "often no better idea of what to do with their money other than to drink it or gamble it away, though their wives might have grander notions."[2] Actually, the wives of some mining men were very much involved in using their families' fortunes to support philanthropic projects in education, thereby performing a public service that has not been sufficiently identified with the history of women in the West.

Of the fifty men in the sample, some information on philanthropy could be found for thirty-six. This undoubtedly is because some kept their philanthropy private or anonymous, while others were not particularly charitable. The wives of others took responsibility for philanthropy. The mining leaders contributed to churches; orphanages; relief organizations for the poor, sick, and disabled; public parks and buildings; and needy friends, business acquaintances, and employees. However, the most consistent pattern was a tendency to support educational institutions and activities, although specific information regarding educational gifts could be found for only twenty-two of the thirty-six philanthropists. When the educational philanthropy of their wives is included the number of cases increases to twenty-five. Despite the limited data, a preference for educational giving was quite clear among the varied contributions of the mining elites and this was notably reflected in gifts to colleges and universities.

In California, Throop Polytechnic Institute in Pasadena received an undisclosed donation from Simeon G. Reed in 1893. Founded two years earlier by Amos G. Throop to teach engineering, arts and crafts, and the humanities, the school was reorganized in 1920 as the now famous scientific research center the California Institute of Technology.[3]

The University of California at Berkeley benefited from the generosity of several mining investors. Darius O. Mills, a leading California banker and Comstock financier, was a regent from 1874 until 1881 and treasurer of the university. In 1881, he gave $75,000 to endow a professorship of intellectual and moral philosophy and civil polity.[4] Mills provided a general guideline as to the nature of the position and the kind of person he hoped would fill it. He also took a personal interest in selecting the first occupant of the endowed chair, Professor George Howison.[5] Although Mills is best remembered in philanthropic circles for the construction after 1888 of a number of "Mills Hotels" in New York City as low-cost apartments for the poor, he also favored the University of California by contributing to and serving as a trustee of its Lick Observatory at Mount Hamilton, one of the first major observatories established in the United States. He was the principal financial donor to the Lick astronomers in Chile after 1900, personally equipping and sending to South America an expedition from the observatory in 1903.[6] William C. Ralston, Mills' partner in organizing the Bank of California and its president from 1873 until his mysterious death in 1875, also was intimately involved with the University of California. Like Mills, he served on the board of regents and was treasurer during the university's early building period in Berkeley.[7] In addition to his policy-making role, Ralston reportedly presented undetermined monetary gifts, a printing press, and men's living quarters to the new facility.[8] The original gymnasium on the Berkeley campus owes its existence to the beneficence of Albert K. P. Harmon, who came to California during the gold rush and later turned a successful mercantile business into profitable investments in a number of Comstock properties in the 1860s. The third building to be constructed on campus, the $15,000 wooden, eight-sided gymnasium was donated to the university in 1878. It served also as an auditorium for musical and dramatic events and commencement exercises since no adequate space for such purposes was then available on campus.[9] Adolph Sutro was another mining magnate who supported the new school at Berkeley. According to his traveling companion and secretary, he helped finance the education of deserving students at the University of California.[10] In 1895, he gave twenty-six acres of San Francisco real estate on Mount Parnassus near Golden Gate Park to the university's affiliated professional schools, including the medical school.[11]

Stanford University was aided by the Henry Newell Scholarships to help finance the education of needy students, especially those from Utah. Newell was an Irish immigrant who began his western career as a meat supplier to miners near Park City, Utah, and later became a director of Utah's Silver King Mining Company. Disappointed by the disinterest of his foster son and grandson in education and intrigued by Leland Stanford's establishment of Stanford University as a memorial to his son, Newell gave $229,996.41 to the university to create an endowed fund. The bequest was Stanford's largest scholarship fund when it was established in 1932.[12] Newell explained: "I have tried to educate two boys who didn't want an education. Now I would like to educate some boys and girls who want an education and can't get it."[13]

The University of Utah at Salt Lake City received a gift from another Utah mining entrepreneur, Enos Wall. This owner of extensive copper properties in Utah's Bingham Canyon and an organizer and director of the Utah Copper Company established and endowed in 1907 the Wall Research Fellowship in metallurgy. Although he withdrew his $500 yearly stipend eight years later to protest the actions of university president Joseph T. Kingsbury toward several faculty members, he nonetheless epitomizes mining leaders who sought to support and honor their profession through their philanthropic gifts and in the process to recognize the value of technological and scientific knowledge in successful mining operations.[14] This is not surprising, since Wall was an authority on the crystalline growth of minerals, a successful inventor of mining and milling machinery, and the discoverer in 1887 of the enormous copper-producing potential of abandoned silver-lead mines in Bingham Canyon, thereby helping to create what has been called "the richest hole on earth."[15]

Jesse Knight was a much better friend of Utah higher education than was Wall. He became known as the "patron saint" of Brigham Young University. His contributions to BYU from 1898 until his death in 1921 amounted to over $500,000 and included the construction of several campus buildings, land, irrigation bonds, and a $200,000 endowed fund to be administered by the Knight Bank and Trust Company.[16] He also was active in the administration of BYU from 1901 until 1921, serving as trustee, chairman of the executive committee of the board of trustees, and vice-president of the board.[17]

The University of Montana received a modest gift from William A. Clark. He preferred other charities to higher education and left relatively little to the state that gave him his fortune. Nonetheless, he provided the law library of the university with a portion of its reference material.[18] He did not follow the advice of John Neill, a local newspaper publisher and political associate, to endow a great university in Helena.[19] A son and heir to the copper fortune was much more generous to higher education. He bequeathed to the University of California at Los Angeles on his death in 1934 the William Andrews Clark, Jr., Memorial Library, which is distinguished for its collection of Dryden, Oscar Wilde, English drama to 1700, early eighteenth-century English literature, and works on the American West and fine printing. He also established a library at the University of Nevada and donated $300,000 for a law building at the University of Virginia.[20]

Colorado College, a small private institution at Colorado Springs, was financially aided by several entrepreneurs who made their fortunes in the Rocky Mountain state. Henry R. Wolcott was a heavy contributor to the college. In 1894, he built an astronomical observatory on campus which also served as a classroom for astronomy courses and as a meeting place for the Colorado College Scientific Society. When he died in 1921, he left $100,000 for the construction of a campus building memorializing his late Congregational minister father, Dr. Samuel Wolcott.[21] Irving Howbert, an early settler and civic leader in Colorado Springs, was a trustee of the college from 1882 to 1922 and trustee emeritus from 1922 to 1934. He also contributed financially to the school. The college recognized his service by awarding him honorary doctoral degrees of laws and letters and by naming a student residence hall in his honor in 1936.[22] Winfield S. Stratton contributed at least $55,000 to Colorado College for fine arts and science buildings. The latter included a museum containing the Henry S. Ward natural history collection, which Stratton purchased for $33,000 in 1900. He even named one of his mining claims the "Professor Lamb" after Henry Lamb, a laboratory assistant in an assaying course he had completed at the college in 1884.[23]

The Colorado State School of Mines at Golden was not neglected by the successful miners. In 1900, Stratton, who served as president of the board of trustees, gave $25,000 to the school for the completion of a metallurgy building. One of the first significant contributions to higher education by a Colorado mining magnate, his gift was commended in a special resolution by the general assembly.[24] Thomas Walsh also showed an interest in the School of Mines. After his son's death in an automobile accident in 1906, Walsh sought to honor his memory by creating the Vinson Walsh Research Fund at the college. The money and scientific equipment donated to the fund were used to support the search for radium.[25]

The University of Denver, which originally began as Colorado Seminary in 1864, benefited from mining money. Dr.

John B. Evans, Colorado's second territorial governor, who had helped to organize Northwestern University in Illinois, founded the seminary by donating land for a campus. However, the institution quickly fell on financially troubled times and discontinued operation in 1867. When it reopened as the University of Denver in 1880, Evans and James B. Grant financially aided the institution. Grant also was a trustee from 1884 until 1904. Even after Evans' death in 1897, the Grant family continued to back the university.[26]

Although the vast majority of the mining leaders who contributed to higher education preferred western institutions, eastern colleges and universities were favored on occasion. John Hays Hammond presented Yale University with the Hammond laboratory of mining and metallurgy in 1905. The gift must have been influenced by his graduation from Yale's Sheffield School of Mining in 1876 and his employment as a professor of mining at Yale in 1902–3. The university also awarded him honorary degrees in 1898 and again in 1927.[27] Wolcott left $25,000 to Yale in 1921.[28] The largest gift to higher education from the mining philanthropists was bequeathed by Joseph R. DeLamar. In his will he left nearly half of his $20 million estate to the Harvard University Medical School, Johns Hopkins University, and the College of Physicians and Surgeons of Columbia University for research into causes and prevention of disease.[29] Thomas Kearns was a member of the board of trustees of the Catholic University of America in Washington, D.C., from 1911 until his death in 1918. Given his generous contributions to Catholic charities, it is likely that he also financially supported the university he helped to govern.[30]

Despite a preference for colleges and universities in their educational philanthropy, the mining leaders did not overlook the elementary and secondary levels. Simeon G. Reed donated several lots to a boys' school in Tacoma, Washington.[31] Henry R. Wolcott financially underwrote his sister's private, college preparatory school for girls, the Miss Wolcott School in Denver. He also took an interest in Denver public school education, creating a prize for the best public reader among female high school pupils.[32] William A. Clark established a fund of $100,000 for kindergarten work, which was donated in the name of his first wife, Katherine Stauffer Clark, to the Kindergarten Association of New York City.[33] Adolph Sutro, who claimed that "charity and kindness are the true religion of the world," gave as one of his earliest philanthropical gifts twenty dollars per month to the Silver Street Kindergarten in San Francisco.[34] The first free kindergarten in California, the school was opened in 1878 by Kate Douglas Wiggin, who trained most of California's early kindergarten teachers and later became a popular writer of children's fiction.[35] Samuel Newhouse, a mining investor who played a substantial part in developing Utah's copper industry as the owner of the Highland Boy Mine in Bingham Canyon, gave $5,000 in 1907 to David McKay, head of Weber State Academy, for the completion of the school's new building. The academy later became Weber State University in Ogden, Utah.[36] Another Utah mine developer, John Q. Packard, constructed a public school building for Eureka, Utah.[37] Thomas Selby, whose Selby Silver and Lead Smelting Works in San Francisco was the first successful reduction plant of its kind in California, liberally supported two private schools in nearby San Mateo County. They were most likely the San Mateo Institute for Young Ladies and St. Matthew's School, locally prominent institutions established between 1860 and 1870.[38]

If educational giving is defined in a broad sense to include donations to libraries and art museums, an even larger number of the mining entrepreneurs were educational philanthropists.[39] John Q. Packard was a prime example. Having made most of his money in Utah mining, he donated $120,000 in the form of a lot and public library building to Salt Lake City in 1904. He also contributed $100,000 in 1905 to build a library in Marysville,

California, where he and a partner, Colonel Edwards Woodruff, had maintained a successful merchandising business from 1849 until 1862, when a flood forced them to sell.[40] In 1900, Thomas Walsh gave a public library to Ouray, Colorado, the mining camp where he had made his fortune.[41] Another Colorado capitalist, David H. Moffat, honored his hometown, Washingtonville, Orange County, New York, and commemorated his grandfather and mother by giving $30,000 to erect the Moffat Library and Hall.[42] Horace Tabor included the donation of books to libraries among his many open-handed charities.[43] In 1880, Adolph Sutro gave $1,000 to the San Francisco Free Public Library for the acquisition of books on mining and related geological sciences.[44] Although William A. Clark did not found a library, he purchased the Major Allen Collection of books for the Montana Historical Society in Helena.[45]

A number of the mining leaders had private art collections, but Clark was one of the few who made his available to the public. He left his extensive collection, valued at $3 million, to the Corcoran Gallery in Washington, D.C., when the Metropolitan Museum of Art in New York City refused to accept his request that the collection should remain intact and be exhibited under his name. Although the paintings were not consistently of high quality, a number were by Titian, Van Dyck, Rembrandt, Hals, and Rubens and there were some outstanding works by Reynolds, Hogarth, Gainsborough, and Raeburn.[46] Like Clark, Darius O. Mills also subsidized the arts. He spent several thousand dollars founding an art gallery in Sacramento, and was a generous contributor to the Metropolitan Museum of Art, serving also as trustee.[47] His former banking partner, William C. Ralston, was a patron of the arts in San Francisco, where he helped to found the San Francisco Art Union in 1865. In 1871, this organization was absorbed by the San Francisco Art Association, of which Mills and other prominent businessmen were members. The association promoted painting, sculpture, and the fine arts through periodic receptions and community exhibitions. In 1874, it opened the California School of Design, the first art school west of the Mississippi. The school was supported by public contributions and private donations, including the former family mansion of Mark Hopkins, the railroad king. In the twentieth century, the institution was renamed the California School of Fine Arts.[48]

Although it is difficult to explain with certainty why some mining magnates were willing to give in behalf of educational and aesthetic activities, a number of factors are significant. By the late nineteenth century, philanthropy clearly had become the responsibility of the wealthy, especially in business. Writing in the 1890s, E. L. Godkin asserted that "the modern philanthropic movement, in particular, has laid heavy burdens on [the rich man]. He is now allowed to have wealth, but the ethical writers and the clergy supervise his expenditure closely. If he does not give freely for charitable objects, or for the support of institutions of beneficence, he is severely criticized. His stewardship is insisted on."[49] Judging by the attitudes of Jesse Knight of Utah, Winfield S. Stratton of Colorado, Albert K. P. Harmon of California, and John Mackay of Nevada, some of the mining men took the doctrine of stewardship seriously. According to Knight's son, Jesse "tried to impress his children that any money we should get should be used wisely, for he thought it was being shown to him for the purpose of doing good... he regarded the matter as a trusted stewardship."[50] Stratton similarly reflected: "I count my money as a gift from the Father of us all, and I am responsible for its administration as a good steward."[51] Harmon declared: "I am not unmindful of the obligations resting upon all good citizens to contribute from their means to charity and to the public good. . . ."[52] According to a contemporary of Mackay, "The thought that controlled him was that wealth in this world is a trust; that the greater the wealth, the more exacting is the obligation to use it righteously."[53]

In 1889, Andrew Carnegie gave the idea of stewardship or trusteeship specific form. In the Gospel of Wealth he called upon the rich to regard their surplus wealth as a public trust to be distributed for benevolent purposes. Among the various appropriate institutions of beneficence, he included universities, astronomical observatories, and library buildings.[54] Prosperous business leaders either followed Carnegie's direction, as demonstrated by the many public libraries bearing his name, or anticipated his suggestion. The University of Chicago was endowed in 1890 by John D. Rockefeller, and such other wealthy businessmen as Ezra Cornell, Johns Hopkins, Cornelius Vanderbilt, Jonas Clark, and James Duke gave generously to educational institutions that now bear their names. Nonetheless, despite substantial gifts and endowments, it is doubtful that contributions by the wealthy exceeded state and national governmental grants to state universities and local colleges in the late nineteenth century. At the same time, a contemporary investigation of private individual benefactions for the period from 1893 to 1903 concluded that a majority of the gifts and bequests went to colleges and universities.[55] A systematic study of the philanthropy of fifty-six predominantly eastern elites drawn from the list of reputed millionaires compiled by the *New York Tribune* in 1892 found that slightly over a third of the donations, or the highest number in any category, were for education in a broad sense. They comprised aid to liberal arts, technical and vocational education, libraries, adult education, and the fine arts.[56] Not only did members of the business community increase their donations to higher education in this period, they assumed positions of leadership in greater number on college and university boards of trustees, displacing clerics who had dominated these governing boards in the antebellum era. Their participation was motivated in part by a desire to reform the traditional classical curriculum in the direction of a more scientific, technical, and business-oriented education.[57]

It is conceivable that Carnegie's Gospel of Wealth, with its emphasis on educational giving and the educational benefactions and administrative roles of established, mostly eastern, business elites, defined the social role for impressionable western mining nouveaux riches to follow. By financially supporting colleges and universities and by serving on their governing boards, socially ambitious mining magnates gained in part the respectability and status they sought. John Hays Hammond, who knew many western mining elites personally and professionally, commented: "I have observed in the case of the nouveaux riches the desire to turn their wealth immediately into some tangible evidence of their improved economic and social status."[58] The noted business historian Thomas C. Cochran claims that "because philanthropic endeavor was above criticism as a path to social distinction, it held particular appeal for the newly rich."[59] Thus, educational philanthropy and service allowed western mining moguls to fulfill the social expectations and obligations of their new status as notable business successes.

It would be an exaggeration, however, to assume that educational giving was entirely a socially conditioned response to outstanding achievement in business. Undoubtedly some of the mining philanthropists gave from a serious personal desire to upgrade the level of learning, especially in the West. Others sought to perpetuate their own reputations or to honor a family member. Still others most likely saw educational support as a convenient way to avoid the pitfall of indiscriminate handouts which could make them vulnerable to continuous appeals for charity from various individuals. Knight's impatience with this problem is reflected in a letter to a Mormon bishop: "I have a request from Sylvia Blanchard for $200.00 with which to build her mother a room. I am not acquainted with her but feel like they are worthy but would rather that it go through the channel of the church and not to be publicly known in it myself for the reason that one tells another and there is no end to the things that are asked for. . . ."[60] Given such circumstances, some of the mining leaders preferred to give to community institutions such as colleges rather than to personal hardship cases.

Gifts to the latter could also undermine self-reliance. Many mining magnates would probably have agreed with John Mackay, whose open-handed generosity included donations to professional panhandlers, unemployed actors, Comstock pensioners, and poor old Irish ladies.[61] In the late 1890s, he confessed: "I want to help people; I like to do it; but when I look back and see the result of trying, I'm ready to swear I'll never give or lend another friend a dollar. . . . It takes the backbone out of a man to have money given to him. . . ."[62] Contributions to libraries, schools, colleges, and universities eliminated the problem since the mining leaders were thereby helping those who were willing to help themselves. As Mills remarked, "I believe in helping those who will help themselves. Then the recipient of a favor, however small it may be, maintains his self-respect."[63]

An analysis of the personal backgrounds of the twenty-two mining men who were known to have supported educational institutions provides additional insight into their motivation. Grammar school was the highest educational level achieved by thirty-six percent, high school by fifty-five percent, and college by nine percent. Included in "grammar schools" are institutions called by that name, as well as district, public, common, and similar schools. In "high school" are included academies and other institutions of similar rank. Counted among grammar school boys are those who had little or no formal education as well as graduates; among high school boys, all those who attended whether graduates or not. The same rule applies to college men. In some cases individuals returned to school after having assumed their first regular jobs. These statistics refer to the educational level reached before going to work for the first time.

It would be useful before drawing conclusions to know the educational levels attained by the general population in the West and by other business leaders who financially supported educational institutions. Nonetheless, given the apparently limited educational experience of the subjects of this study, it is possible that they were more aware of the importance of providing the less fortunate with access to schools, colleges, universities, libraries, and art galleries than were college-educated entrepreneurs who might have been more inclined to take such opportunities for granted. Having been denied extensive education as youth, these men perhaps could more keenly appreciate its value for success and happiness. In fact, it is instructive to note that such poor boys with very little formal education as Jesse Knight, Henry Newell, and Joseph R. DeLamar were among the most generous of the group in their support for colleges and universities. Also, the more practical orientation of post–Civil War institutions of higher education, featuring greater attention to management or administration and other characteristics of the modern university, might have made them easier to support for less well educated businessmen than institutions emphasizing the traditional classical form of collegiate training. Indeed, according to Merle Curti and Roderick Nash, benefactors of practical higher education from the business community in the nineteenth century were, with few exceptions, not college graduates. "Most of them had not needed the traditional curriculum to make a success in their professions and consequently had little desire to perpetuate what they considered a sterile tradition."[64]

The educational charity of some mining leaders also was shaped by the interests of their wives, who became philanthropists in their own right. Such a woman was Phoebe Apperson Hearst. A former Missouri school teacher, in 1862 she married George Hearst, a man more than twice her age. When he died in 1891, Phoebe inherited his entire estate and became a majority owner of the Homestake Mining Company. As such she took a special interest in Lead, South Dakota. In 1894, she gave a complete library to the community as a Christmas gift. In addition to regular services, the library provided books, magazines, and newspapers in various foreign languages to accommodate the needs of the immigrant workers and their families. Hearst continued to financially support the library until her death in 1919.[65] She also created the Hearst Free Library to benefit the mining families of Anaconda, Montana, eventually giving the library to the city in 1904.[66]

Even before her husband's death, Hearst had become intimately involved with the early kindergarten movement. In San Francisco she helped Sarah B. Cooper, known for her educational work with slum children in the crime-laden Barbary Coast District, maintain free kindergartens long before they were officially incorporated into the public school system. In 1883, she financed a building that housed several kindergarten classes and Cooper's Golden Gate Kindergarten Association, for which Hearst served as honorary president. Although she was a major financial donor, her role was not unique because the San Francisco kindergarten movement received support and leadership from other members of the city's social and economic elite who hoped "to instill in the children of the poor a commitment to work and morality which would make them productive citizens rather than a threat to society."[67] According to Carol Roland's excellent study on the California kindergarten movement, by 1906, Hearst had contributed over $80,000 to San Francisco kindergartens. She eventually established a trust fund for the Golden Gate Kindergarten Association which still exists and provides support for the Phoebe Hearst Pre-School Learning Center in the city.[68]

Hearst's work in behalf of early childhood education was not confined to San Francisco. In the nation's capital she founded kindergartens for white and black children in addition to the National Cathedral School for Girls, which was presented to the district Episcopal diocese in 1896 as a means of providing a Christian education in particular for the daughters of federal government employees. She also helped establish the Columbia Kindergarten Association, becoming its first president in 1893. It was organized to pressure Congress into enacting legislation bringing kindergartens into the public schools of the federal district.[69] At the same time, Hearst did not neglect the mining camps where her husband had prospered. After 1901, she was the sole support of the new kindergarten in Lead, hiring teachers who were graduates of the Phoebe Hearst Kindergarten Training School she had built in 1897 in Washington, D.C. The kindergarten served the needs of immigrant children by introducing them to the English language and American culture, thereby making them more competitive in the public schools with American-born children. In short, Hearst promoted kindergarten training as a first step or head start in their Americanization. She further helped the public schools in Lead by purchasing construction bonds at a low rate of interest.[70]

The later years of her life were identified especially with the state university at Berkeley. She was the first female regent of the University of California from 1897 until her death and one of the most generous benefactors in its history. Her many contributions included a museum of anthropology, and, as a testimony to her late husband, the Hearst Memorial Mining Building, which was constructed at a cost of $645,000 and presented to the university in 1908. The additional expense of equipping and maintaining the building increased the cost to $800,000. In addition, she gave the university the women's social and athletic center known as Hearst Hall. Destroyed by fire in 1922, it had been personally designed by the gifted Berkeley architect Bernard Maybeck.[71] Hearst also helped to establish a department of anthropology at Berkeley, which included such renowned scholars as Alfred Kroeber. She financed university archaeological expeditions to Italy, Greece, Mexico, South America, and Egypt. In

1905, while visiting the latter country, she even set up temporary living quarters in a small hospital at the diggings.[72] Perhaps most significantly, she underwrote at a cost of $200,000 the international competition to determine a comprehensive architectural plan for the Berkeley campus which attracted over one hundred contestants. The winning design was submitted by Henri Jean Emile Bernard of Paris, who received the $50,000 first prize. Less spectacular gifts included books, manuscripts, paintings, prints, salaries for professors, laboratory equipment, concerts for students by the Metropolitan Opera Company, basketball and tennis courts, and swimming pools.[73] At the same time, Hearst took a personal interest in the Berkeley students, regularly meeting with the female students in particular, for whom she had created over twenty scholarships and a work-study program. Those who needed jobs to defray the cost of their education found employment as seamstresses in the Hearst Domestic Industries, established at the university in 1900.[74] At her home, the Hacienda del Pozo de Verona, in Pleasanton, California, she entertained the entire senior class each spring. Upon Hearst's death, the president of the University, Benjamin Ide Wheeler, whom she as regent had helped to recruit from Cornell University, eulogized her as "a gentlewoman and public servant—a blessing to her day and generation."[75] Although William Randolph Hearst, her famous son, was named the chief beneficiary of his mother's $11 million estate, the University of California was the only institution financially provided for in her will. It received her extensive art collection and $60,000 in trust to continue the student scholarship program. Her total contributions to the university over the years exceeded $1.5 million.[76] During her life, she had given away $21 million to educational and other charities. One cannot discount her genuine interest in using her husband's mining fortune, conservatively estimated at $18 million in 1891, for humanitarian causes.[77]

Although Phoebe Hearst was the most notable philanthropist among the wives of the mining magnates, others drew upon their husbands' resources to promote education. Simeon G. Reed had instructed his wife to devote part of his estate "to the cultivation, illustration or development of the fine arts of said city of Portland, or to some other suitable purpose, which shall be of permanent value and contribution to the beauty of the city and to the intelligence, prosperity and happiness of its inhabitants."[78] When Amanda Reed died in 1904, nine years after her husband, she left $3 million toward the founding of Reed College in Portland, which opened in 1911.[79] Amanda Knight, who shared her husband's interest in Brigham Young University, established the Amanda Knight scholarships for deserving students. In 1898, she donated seven and a half acres of land on Temple Hill near the campus to the Provo Fourth Ward of the Church of Latter-Day Saints. Known as Raymond Park in honor of Knight's eldest son, the land subsequently was transferred to the University with her blessing in 1904.[80] Some of the millions made in the 1870s by John Mackay on the Comstock Lode were given by his wife, Marie, and son, Clarence, to the University of Nevada after John's death in 1902. The money was used to build the Mackay School of Mines, the Mackay Science Building, Mackay Stadium, and the Mackay Statue by Gutzon Borghum.[81] Although Marie Mackay was much more at home in European society than in a crude mining town like Virginia City, Nevada, she did not forget the state that had given her the means to pursue her social ambitions as a continental hostess and bon vivant.[82] Alice Hale Hill was active in making kindergarten education available to students in Denver. Married to Nathaniel P. Hill, she became a prominent figure in Denver society in the 1880s and 1890s, but, as her diary indicates, she also invested time and presumably money in the local kindergarten movement: "1890 (Thurs.) Kindergarten met here—examined plan for the Building—Discovered ways of raising money."[83] "1891 (Thurs.) In A.M. I went to State Teachers Assoc. . . . Interesting paper from Prof. Parker of Ft. Collins and discussion afterwards—The entire sentiment was in favor of attaching the Kindergarten to the public schools. . . ."[84] Although the mining wives gave time and money to a variety of charities—such as churches, ladies relief societies, hospitals, and the Young Women's Christian Association—Phoebe Hearst, Amanda Reed, Amanda Knight, Marie Mackay, and Alice Hill were best known for their support of education.

The most salient pattern in the educational philanthropy of the mining families was provincialism. Although some contributed to eastern schools, universities, libraries, and art galleries, the overwhelming majority preferred to support western institutions. Apparently they sought to give something back to the region that had made them prosperous, and they thereby became important western community builders. This point requires emphasis because the mining moguls, like other nouveaux riches of the period, have been perceived popularly as lacking a sense of social propriety and responsibility. The sensationalistic social climbing of such mining wives as the "Unsinkable" Molly Brown, Marie Mackay, and Ida Newhouse; the scandalous separation of Horace Tabor from his long-suffering first wife in order to marry the young, beautiful divorcee Lizzie "Baby Doe" McCourt; Comstock financier William C. Sharon's lengthy legal battle with his mistress, Sarah Althea Hill, over a disputed marriage contract; the vast expenditures of those such as Thomas Walsh, whose lavish banquets abroad in 1900 made him the talk of Paris; and such opulent mansions and country estates as James Flood's incredible Linden Towers, known as "Flood's wedding cake," in Menlo Park, California, have created a stereotype of extravagant, frivolous, and even bizarre personal indulgence.[85] While such an image cannot be totally discounted, the philanthropy of the mining leaders, whether for educational or other causes, reveals a public-spirited profile that has not been sufficiently recognized.

NOTES

1. The standard criticism of the nineteenth-century business leaders is contained in Matthew Josephson, *The Robber Barons: The Great American Capitalists, 1861–1901* (New York: Harcourt, Brace and World, 1962 [orig. pub. 1934), while a balanced but positive view of a leading entrepreneur can be found in Allan Nevins, *Study in Power: John D. Rockefeller, Industrialist and Philanthropist*, 2 vols. (New York: Charles Scribner's Sons, 1953). The continuing "robber baron" controversy can be traced in Hal Bridges, "The Robber Baron Concept in American History," *Business History Review* 33 (Spring 1958): 1–13; David Chalmers, "From Robber Barons to Industrial Statesmen: Standard Oil and the Business Historians," *American Journal of Economics and Sociology* 20 (October 1960): 47–58; Edward C. Kirkland, "The Robber Barons Revisited," *American Historical Review* 66 (October 1960): 68–73; Thomas C. Cochran, "The Legend of the Robber Barons," *Pennsylvania Magazine of History and Biography* 76 (July 1950): 307–21; and Fritz Redlich, "The Business Leader as a 'Daimonic' Figure," *American Journal of Economics and Sociology* 20 (January-April 1953): 163–78, 289–99.

2. Joseph J. Thorndike, Jr., *The Very Rich: A History of Wealth* (New York: American Heritage Publishing Co., Bonanza Books, 1981), 77.

3. F. J. Polley to Reed, 28 June 1893, Reed Papers, 35:202.

4. *Alta California* (San Francisco), 10 July 1881; W. F. Swasey, *The Early Days and Men of California* (Oakland: Pacific Press Publishing Co., 1891), 344; William W. Ferrier, *Origin and Development of the University of California* (Berkeley: Sather Gate Book Shop, 1930), 434–35; Johnson et al., eds., *Dictionary of American Biography*, 13:6–7; and *Who's Who in America* (Chicago: A. N. Marquis and Co., 1908–9), 5:1308. A contemporary sketch of Mills can also be found in Samuel E. Moffett, "Darius Ogden Mills," *The Cosmopolitan* 33 (May–October 1902): 289–92.

5. Mills to the University of California Board of Regents, 7 July 1881, Mills folder, George Holmes Howison Papers, 1862–1917, Bancroft Library, University of California, Berkeley. See also letters, Mills to Howison, 1883–1901, ibid.

6. Johnson et al., eds., *Dictionary of American Biography*, 13:6–7; and Verne A. Stadtman, *The University of California, 1868–1968* (New York: McGraw-Hill, 1970), 204. Commentary on the hotels is in Goodwin, *As I Remember Them*, 59–60.

7. For a description of Ralston's role in the university's early history, see Lavender, *Nothing Seemed Impossible*, 343–47; and Stadtman, *The University of California*, 39, 42, 48.

8. Tilton, *William Chapman Ralston*, 315–16; and Julian Dana, *The Man Who Built San Francisco: A Study of Ralston's Journey with Banners* (New York: Macmillan Co., 1936), 380–81.

9. See Harmon's obituary in the *Oakland Enquirer* (Oakland, California), 30 April 1896, and a description of the gymnasium in the *Alta California* (San Francisco), 22 October 1878. For a photograph of the original gymnasium, see Russell Schoch, "Gym Dandy," *California Monthly* 95 (October 1984): 27. The original building was torn down in 1933 and replaced with a more modern facility. See Stadtman, *The University of California*, 110. A brief biographical sketch of Harmon is in Angel, *History of Nevada*, unnumbered pages between 116 and 117. Harmon also served as a trustee of Mills College in Oakland, California, known as Mills Seminary until 1885. See *Oakland Enquirer* obit. above.

10. Unsworth, "Memoir," II, 49–51.

11. Stewart, *Adolph Sutro*, 207, 213. For the donation of land to the University of California, see University of California Regents, *Annual Report of the Secretary to the Board of Regents of the University of California* [for the Year Ending June 30, 1896. (Sacramento: State Printing Office, 1896), 85–92.

12. *Henry Newell: An Account of the Events in His Life Leading to the Establishment of the Henry Newell Scholarships at Stanford University* (Stanford: Stanford University Press, 1954), 3–15.

13. Ibid., 10.

14. G. W. Barrett, "Enos Andrew Wall, Mine Superintendent and Inventor," *Idaho Yesterdays* 15 (Spring 1971): 30–31; and *Mining and Scientific Press* (San Francisco), 10 July 1920.

15. For Wall's mining career, see G. W. Barrett, "Colonel E. A. Wall: Mines, Miners, and Mormons," *Idaho Yesterdays* 14 (Fall 1970): 3–11; and *National Cyclopedia of American Biography*, 20:145. Utah's Bingham Canyon is discussed in Leonard J. Arrington and Gary B. Hansen, *The Richest Hole on Earth: A History of the Bingham Copper Mine* (Logan: Utah State University Press, 1963).

16. Jesse William Knight, *The Jesse Knight Family, His Forebears and Family* (Salt Lake City: Deseret News Press, 1940), 73, 89–92, 96; and Jesse Knight Folder, Knight Papers.

17. Knight, *The Jesse Knight Family*, 98.

18. Mary M. Farrell, "William Andrews Clark" (M.A. thesis, University of Washington, 1933), 10. William D. Mangam, *The Clarks, An American Phenomenon* (New York: Silver Bow Press, 1941), 90–92, contains a brief reporting of Clark's charities.

19. Malone, *The Battle for Butte*, 111, 200.

20. Brief descriptions of the library at UCLA are available in James High, "William Andrews Clark, Westerner: An Interpretive Vignette," *Arizona and the West* 2 (Autumn 1960): 245; and in James D. Hart, *A Companion to California* (New York: Oxford University Press, 1978), 85. See Mangam, *The Clarks*, 197–200, for his other charities, including the establishment of the Los Angeles Philharmonic Orchestra.

21. Wolcott, "Dictation," no pagination; *Denver News*, 1 June 1921; *Denver Post*, 24 June 1921; and J. Juan Reid, *Colorado College: The First Century, 1874–1974* (Colorado Springs: Colorado College, 1979), 47, 93.

22. Howbert, *Memories of a Lifetime*, 287–90; *Denver Post*, 21 December 1934; Reid, *Colorado College*, 140; and a letter, George V. Fagan, Head Librarian, Colorado College, to the author, 5 August 1982, regarding Howbert.

23. Waters, *Midas of the Rockies*, 250; Smiley, *History of Denver*, 566; Reid, *Colorado College*, 42, 60; and Dawson, Scrapbooks, Mining, II, 23:521.

24. Smiley, *History of Denver*, 566; Waters, *Midas of the Rockies*, 155; and Michael McGiffert, *The Higher Learning in Colorado: An Historical Study, 1860–1940* (Denver: Sage Books, 1964), 24. "The best thing that ever happened to this school was its discovery by Winfield S. Stratton. When he consented to become president of its board of trustees [in 1899 the entire faculty must have sung a quiet Te Deum." See *Colorado Transcript* (Golden, Colorado), 6 February 1901.

25. *Denver Republican*, 9 April 1910; and Will Ferrill, ed., *Sketches of Colorado* (Denver: Western Press Bureau Co., 1911), 1:329. For an analysis of Walsh's other personal and family expenditures, see Richard H. Peterson, "Thomas F. Walsh: The Lifestyle of a Colorado Mining Magnate, 1896–1910," *Red River Valley Historical Review* 6 (Fall 1981): 53–63.

26. Lyle W. Dorsett, *The Queen City: A History of Denver* (Boulder: Pruett Publishing Co., 1977), 47–48, 80, 89; Harry E. Kelsey, Jr., *Frontier Capitalist: The Life of John Evans* (Denver: State Historical Society of Colorado, 1969), 219–25; McGiffert, *The Higher Learning in Colorado*, 26; and Johnson et al., eds., *Dictionary of American Biography*, 7:488–89.

27. *National Cyclopedia of American Biography*, 26:45–46.

28. *Denver Post*, 24 June 1921.

29. Ibid., 28:257–58; and *New York Times*, 10 December 1918.

30. Larsen, "The Life of Thomas Kearns," 150–51. See also the *Deseret Evening News* (Salt Lake City), 19 October 1918; Whitney, *History of Utah*, 4:697; and Helen M. Crampton, "A Sketch of the Kearns Family," 1961, Utah State Historical Society, Salt Lake City, 2.

31. T. B. Wallace to Reed, 27 October 1885, Reed Papers, 18:166–67. Wallace was Reed's Tacoma land agent.

32. *Denver Post*, 1 June 1921; and Dorsett, *The Queen City*, 89, 163.

33. Mangam, *The Clarks*, 91.

34. Stewart, *Adolph Sutro*, 177, 192.

35. See Doyce B. Nunis, Jr., "Kate Douglas Wiggin: Pioneer in California Kindergarten Education," *California Historical Society Quarterly* 41 (December 1962): 291–307. The earliest private, tuition-based, California kindergarten was opened in San Francisco in 1863. However, it was not until 1913 that the state legislature permitted the incorporation of kindergartens into the public school system if desired by a sufficient number of parents in local school districts. See William W. Ferrier, *Ninety Years of Education in California, 1846–1936: A Presentation of Educational Movements and Their Outcome in Education Today* (Berkeley: Sather Gate Book Shop, 1937), 163–64, 174.

36. *Deseret Evening News*, 20 March 1907. For Newhouse's career, see Hynda Rudd, "Samuel Newhouse: Utah Mining Magnate and Land Developer," *Western States Jewish Historical Quarterly* 11 (July 1979): 291–307.

37. Goodwin, *As I Remember Them*, 330–31.

38. *Alta California*, 10 June 1875; and Ferrier, *Ninety Years of Education in California*, 141–42. Selby was also a trustee of what the *Alta California* called the "City College" in San Francisco. This most likely is a reference to the first normal school in San Francisco which began training teachers in 1862 and later moved to San Jose, to become San Jose State College (now University). See *Alta California*, 8 September 1869; and Doris Muscatine, *Old San Francisco: The Biography of a City from Early Days to the Earthquake* (New York: G. P. Putnam's Sons, 1975), 181.

39. Construction and funding of opera houses, such as those built by Horace Tabor in Leadville and Denver, and theaters, such as William C. Ralston's California Theater in San Francisco, were not considered as philanthropy. Although they properly belong to the arts and reflect the civic

consciousness of their builders, they were designed as profit-making enterprises.

40. Guinn, *History of the State of California*, 360–63.

41. *Denver Republican*, 9 April 1910; and Ferrill, ed., *Sketches of Colorado*, 1:329.

42. Henry R. Wolcott, "Dictation and Related Biographical Material for David H. Moffat," 7; Walter Chessman, ibid., 5–6; and Akers, "David H. Moffat and His Home Town," 212. Moffat was also the first treasurer and an original trustee of Colorado Seminary, later the University of Denver. See Kelsey, *Frontier Capitalist*, 221.

43. Smith, *Horace Tabor*, 253–54.

44. See Sutro's letter to Andrew J. Moulder, secretary of the library, in the *Alta California*, 1 January 1881. Moulder was California's third superintendent of public instruction, from 1856 until 1862. His career is discussed in James E. Myers, "The Educational Work of Andrew Jackson Moulder" (Ph.D. diss., University of California, Berkeley, 1961).

45. Farrell, "William Andrews Clark," 10.

46. Mangam, *The Clarks*, 91–92; and Johnson et al., eds., *Dictionary of American Biography*, 4:145.

47. Johnson et al., eds., *Dictionary of American Biography*, 13:7; Hall, ed., *America's Successful Men of Affairs*, 1:444; and "American Millionaires and Their Public Gifts," *Review of Reviews* 7 (February 1893): 57.

48. Charles C. Hoag, ed., *The San Francisco Blue Book* (San Francisco: Charles C. Hoag, 1902), 335–39; Lavender, *Nothing Seemed Impossible*, 337–38; Hart, *A Companion to California*, 374–75; and Muscatine, *Old San Francisco*, 156.

49. E. L. Godkin, "The Expenditure of Rich Men," in *The Nation Transformed, The Creation of an Industrial Society*, ed. Sigmund Diamond (New York: George Braziller, 1963), 120. (Orig. pub. in *Scribner's Magazine* 20 [1896]: 495–501.)

50. Knight, *The Jesse Knight Family*, 39–40.

51. Waters, *Midas of the Rockies*, 191. See also the *Denver News*, 15 September 1902.

52. *Oakland Enquirer*, 7 May 1896.

53. Goodwin, *As I Remember Them*, 169.

54. See Andrew Carnegie, "Wealth," *North American Review* 148 (1889): 657–62; and "The Best Fields for Philanthropy," ibid., 149 (1889): 687–96.

55. George J. Hagar, "Magnitude of American Benefactions," *Review of Reviews* 29 (April 1904): 464–65.

56. Merle Curti, Judith Green, and Roderick Nash, "Anatomy of Giving: Millionaires in the Late 19th Century," *American Quarterly* 15 (1963): 425–26. Curti also discovered that donations in relief of the poor, sick, and aged constituted a third of the total. Religious organizations received about a quarter of the donations, while less than a tenth went to civic improvement, namely to public buildings, monuments, and parks.

57. For the role of business leaders in higher educational reform and the creation of the modern university, see Edward C. Kirkland, *Dream and Thought in the Business Community, 1860–1900* (Chicago: Quadrangle Books, 1964), 83–113; and Merle Curti and Roderick Nash, *Philanthropy in the Shaping of American Higher Education* (New Brunswick: Rutgers University Press, 1965), 107–31. At times, as in the 1890s, the participation of business patrons in shaping educational policy led to infringements of academic freedom, especially when administrators or professors were dismissed for embracing political or economic views that were perceived as a threat to big business or the conservative social order. The historically significant case involving the dismissal of Professor Edward A. Ross of Stanford University in 1900 exemplifies this confrontation. For an analysis of this and related issues, see ibid., 131–35; and Richard Hofstadter and Walter P. Metzger, *The Development of Academic Freedom in the United States* (New York: Columbia University Press, 1955), 413–67.

58. Hammond, *Autobiography*, 2:504. "A trusteeship in a large university became, along with a listing in the Social Register, a token of business prominence and of pecuniary qualification" (Hofstadter and Metzger, *Development of Academic Freedom*, 415).

59. Thomas C. Cochran, *Business in American Life: A History* (New York: McGraw-Hill, Inc., 1974), 238.

60. Knight to Bishop O. B. Huntington, 1 July 1909, box 49, file H., Knight Papers. For evidence of the problem described by Knight, see begging letters sent to John W. Mackay, *San Francisco Call*, 4 November 1883.

61. For an indication of the various people Mackay helped to support, see *Alta California*, 10 January 1881; Goodwin, *As I Remember Them*, 165–67; and Oscar Lewis, *Silver Kings: The Lives and Times of Mackay, Fair, Flood, and O'Brien, Lords of the Nevada Comstock Lode* (New York: Alfred A. Knopf, 1967 [orig. pub. 1947), 55–62.

62. Quoted in Lewis, *Silver Kings*, 62. James C. Flood similarly reflected that "to dole out unearned cash destroyed not only the recipient's initiative but his self-respect as well. . . ." (ibid., 235).

63. Moffett, "Darius Ogden Mills," 291–92.

64. Curti and Nash, *Philanthropy*, 61–62.

65. W. A. Swanberg, *Citizen Hearst: A Biography of William Randolph Hearst* (New York: Charles Scribner's Sons, 1961), 4–6; Hearst, *The Way It Was*, 5–18; and Joseph H. Cash, *Working the Homestake* (Ames: Iowa State University Press, 1973), 70–72.

66. Fremont Older and Cora Older, *George Hearst, California Pioneer* (Los Angeles: Westernlore Press, 1966 [orig. pub. 1933), 166; and Malone, *The Battle for Butte*, 78.

67. See Carol Roland, "The California Kindergarten Movement: A Study in Class and Social Feminism" (Ph.D. diss., University of California, Riverside, 1980), 6–7, 13–14, 21–22, 98; and the *San Francisco Examiner*, 15 April 1919. See also letters requesting financial aid or expressing gratitude: Sarah Cooper to Phoebe Hearst, 23 July 1888, 12 August 1889, 16 December 1892, and 2 March 1893, Incoming Correspondence, Phoebe Apperson Hearst Papers, Bancroft Library, University of California, Berkeley (hereafter cited as the PAH Papers). In 1887, Cooper named a kindergarten in honor of Phoebe Hearst. This common practice was designed to raise money from primary donors. See Roland, "The California Kindergarten Movement," 98.

68. Ibid., 97, 122 n. 20.

69. *San Francisco Examiner*, 14, 16 April 1919; Johnson et al., eds., *Dictionary of American Biography*, 8:488; *National Cyclopedia of American Biography*, 25:322; Elizabeth D. Ross, *The Kindergarten Crusade: The Establishment of Preschool Education in the United States* (Athens: Ohio University Press, 1976), 25, 88; Swanberg, *Citizen Hearst*, 181; and Older, *George Hearst*, 210.

70. Cash, *Working the Homestake*, 72–73; and Ross, *The Kindergarten Crusade*, 25. Carton 9, PAH Papers, also contains information on the Hearst Free Library and Kindergarten in Lead, South Dakota.

71. A complete report of Phoebe Hearst's individual benefactions to the university can be found in "A Gracious Lady Helped the Young," *Oakland Tribune*, 27 April 1969. See also Johnson et al., eds., *Dictionary of American Biography*, 8:488–89; *San Francisco Examiner*, 14 April 1919; Swanberg, *Citizen Hearst*, 272; Stadtman, *The University of California*, 86,

119–20, 137, 160, 177, 191; and the *Los Angeles Herald-Examiner*, 26 September 1976. For a recent concise overview of Maybeck's career, see Georgia Sommers Wright, "Architect's Mind in Flux: Maybeck's *Beaux-Arts* Triumphs, Trials, and Evolution at Mills College," *The Californians* 2 (July/August 1984): 31–38.

72. For Phoebe Hearst's role in the development of the Department of Anthropology and university archaeological expeditions, see Timothy H. H. Thoresen, "Paying the Piper and Calling the Tune: The Beginnings of Academic Anthropology in California," *Journal of the History of the Behavioral Sciences* 11 (July 1975): 257–75; and Edward T. James et al., eds., *Notable American Women, 1607–1950* (Cambridge, Mass.: Harvard University Press, 1971), 2:171. The Hearst biographical essay was written by Rodman W. Paul.

73. *San Francisco Examiner*, 14 April 1919; *National Cyclopedia of American Biography*, 25:322; and Cora [Mrs. Fremont] Older, *William Randolph Hearst, American* (New York: D. Appleton-Century Co., 1936), 529. Also, regarding the architectural plan for the University of California, see letters, Bernard Maybeck to Phoebe Hearst, 1896–99, Incoming Correspondence, PAH Papers.

74. *San Francisco Examiner*, 14 April 1919; *Oakland Tribune*, 27 April 1969. See also letters written by Hearst Domestic Industries students, 1 folder, carton 1 and letters, Amanda Hicks (director) to Phoebe Hearst, 1900–1905, Hearst Domestic Industries file, Incoming Correspondence, PAH Papers. According to the *New York Times*, 28 May 1904, Hearst withdrew her financial support from this organization and several University of California student clubs in 1904.

75. Quoted in the *Oakland Tribune*, 27 April 1969.

76. Ibid.; *San Francisco Examiner*, 14 April 1919; and *New York Times*, 18 April 1919. Phoebe Hearst's estate was appraised at $11,012,850. According to the California State Controller, the state would thus collect an inheritance tax of $1,056,845, of which her son as principal heir would be required to pay $949,101. See *New York Times*, 26 August 1923.

77. Cash, *Working the Homestake*, 72–74; and Swanberg, *Citizen Hearst*, 321. The philanthropic activities of Phoebe Hearst can be traced in Winifred B. Bonfils, *The Life and Personality of Phoebe Apperson Hearst* (San Francisco: John Henry Nash, 1928), however, since it was authorized by her son William Randolph and is badly written, it should be used with care. See also the *San Francisco Examiner*, 14–16 April 1919; and appropri-

ate material in the PAH Papers. The latter contains portions of an unpublished biography with related notes written by Adele S. Brooks.

78. *Reed College Record* (Portland, Oregon), May 1911, 2:6.

79. Ibid.

80. See p. 15 of a manuscript biography of Amanda (Mrs. Jesse) Knight by her daughter in box 168 of the Knight Papers; Knight, *The Jesse Knight Family*, 103–4; and Keith L. Smith, "A History of the Brigham Young University—the Early Years, 1875–1921" (Ph.D. diss., Brigham Young University, 1972), 76–77.

81. Ethel Manter, *Rocket of the Comstock: The Story of John William Mackay* (Caldwell, Idaho: Caxton Printers, 1950), 248; and Lamar, ed., *The Reader's Encyclopedia of the American West*, 697–98.

82. Marie Mackay's social triumphs in Paris and London are recalled in Lewis, *Silver Kings*, 86–101.

83. Notebook 5, p. 51, folder 6, Hill Papers.

84. Ibid., 366.

85. The social activities of Margaret (Molly) Tobin Brown, the wife of Leadville miner James J. Brown, are described in Lucius Beebe, *The Big Spenders* (New York: Simon and Schuster, 1967), 227–34. Horace Tabor's divorce from Augusta and marriage to Elizabeth "Baby" McCourt Doe is examined in Smith, *Horace Tabor*, 175, 211–12, 217–18, 224, 227–29, 231–32, 246–47. Like Molly Brown, Baby Doe has been the subject of books and plays. The American opera *The Ballad of Baby Doe* by Douglas Moore debuted in Central City, Colorado, in 1956. Ida Newhouse's entry into English high society is considered briefly in Rudd, "Samuel Newhouse," 304. The unusual events resulting from the Sharon case are discussed in Milton S. Gould, *A Cast of Hawks* (La Jolla, Calif.: Copley Books, 1985), 169–337. Thomas Walsh's expensive lifestyle is revealed in McLean, *Father Struck It Rich*, 69, 78–79, 91–94, 155–57, 171, 191. Flood's palatial country estate is described in Lewis, *Silver Kings*, 227–29; Gray Brechin, "A Little Place in the Country," *California Monthly* 90 (October 1979): 12–14; and the *San Francisco Call*, 11 January 1880.

3

The Mansion-Building Mania

ALTHOUGH A NUMBER OF MINING ELITES responded positively to the obligations of public service, especially in educational philanthropy, it is equally obvious that the vast majority spent much time and money in private and family self-indulgence. What the contemporary social critic and economist Thorstein Veblen called "conspicuous consumption" aptly characterizes for the most part the personal lifestyle of leading mining entrepreneurs. The big house or Victorian mansion was the centerpiece of luxurious living for virtually all of them. Indeed, to build a fine house has been an impulse of the rich throughout history. Located in an exclusive area of a city such as San Francisco's Nob Hill, Denver's Capitol Hill, or New York's Fifth Avenue or in a summer resort setting such as Newport, Rhode Island, or Long Island's north shore, the grandiose residence symbolized its owner's great wealth and became the site of sumptuous banquets and costume balls and an appropriate arena in which to exhibit an expensive art collection or extensive library. Whether motivated by their own desires, the demands of socially ambitious family members, or the need of the newly rich to define their status as wealthy citizens by the standards of established eastern business society, western mining magnates, like nouveaux riches in other parts of the country, were obsessed with building, decorating, and displaying their stately homes during the great mansion-building era between the Civil War and World War I.

Andrew Carnegie cautioned his fellow millionaires that it is "the duty of the man of Wealth: First to set an example of modest, unostentatious living, shunning display or extravagance."[1] However, even the prudent, sober Carnegie owned a substantial home on fashionable Fifth Avenue in New York City and a mammoth castle in his native Scotland on an estate of 32,000 acres.[2] Most western mining magnates followed the familiar pattern of building a big house resembling the palatial residences of national business leaders. For example, according to architectural historian Harold Kirker, "excepting their wooden construction, the building projects of the [California and Nevada] railroad and silver kings were similar to those of their kind everywhere in the United States in the era of the 'Great Barbecue.'"[3]

The mansions on San Francisco's Nob Hill epitomized the architectural exhibitionism of business executives in the Gilded Age. The first was owned by the lawyer and mining investor James B. Haggin. Completed in 1873, the Haggin residence consisted of two full floors, a stylish mansard roof, and sixty-one rooms, not including three conservatories and many bathrooms and room-size closets. Haggin's house initiated a mansion-building spree on Nob Hill as California millionaires avoided such older, formerly genteel neighborhoods as Rincon Hill, where mining entrepreneurs or financiers such as Lloyd Tevis, William C. Ralston, and Thomas Selby had built their homes. Railroad barons David Colton, Leland Stanford, Charles Crocker, and Mark Hopkins surpassed the Haggin dwelling as they seemingly tried to outdo each other in the size, grandeur, and expense of their respective hilltop palaces. Nob Hill was also the site of the extravagant residences of James Fair and James Flood. Erected in 1886, Flood's "incongruous-looking brownstone pile" was the only mansion not built of redwood. It was also the only Nob Hill town house to survive the 1906 earthquake and fire.[4]

The Comstock mining mogul is better known for the previously mentioned great white mansion on the peninsula south of San Francisco in Menlo Park, San Mateo County. With its mass of turrets and gables and a 150-foot tower, Linden Towers was probably the most opulent country home built in late–nineteenth century California. Completed in 1881, the residence allegedly contained a sample of every known kind of wood. The interior appointments included marble mantlepieces, numerous stained-glass windows, ceiling frescoes painted by Italian artists, a library, dining and billiard rooms, a music conservatory, dancing and smoking rooms, and a drawing-room finished in carved walnut with walls of embossed velvet. The landscaping featured artificial lakes, game preserves, fountains, bronze statuary, and flower gardens.[5]

Flood was not the first mining king to find the San Francisco peninsula a suitable site for mansion building. In 1864, William C. Ralston acquired a sizable country estate in Belmont, San Mateo County, where he expanded an existing villa into an ornate, eighty-eight room Victorian residence with columned doorways, great glass sliding doors, crystal chandeliers, skylit foyers, parquetry floors, and other elegant interior features. A visitor left this description: "This was really a palace, with vast parlors and drawing rooms, marvelous in amplitude. In the music room there was a pipe organ. The house was superbly furnished; carpets and rugs from India and Persia, everything the most luxurious, much of the furniture brought from Europe or the Orient."[6] To the east of the house were a bowling alley and a tile-roofed Turkish bath. Ralston also added an expensive gasworks to illuminate his country seat. After his death in 1875, the estate was acquired by his former business associate and Comstock mining ally, William C. Sharon, who improved it with the addition of a dam and reservoir in the surrounding hills to provide irrigation for extensive trees and gardens.[7]

Darius O. Mills also found San Mateo County a desirable place for country living. In 1860, he became the owner of 1,500 acres on a former Mexican rancho. In 1869, he completed construction of a three-storied, towered, wooden, French mansard house of forty rooms, which he named Millbrae. The grounds featured a botanical garden of exotic plants and trees imported from around the world and a large conservatory for tropical vegetation. In the early 1870s, Chinese laborers reclaimed adjacent marshland which Mills used as pasture for his dairy cattle. Although he later made his home in New York City, he continued to keep Millbrae as a summer home where, like other mining entrepreneurs on the peninsula, he could live like a country gentleman.[8]

In the village of San Mateo, another early Ralston business associate and Comstock mine developer, Alvinza Hayward, settled on seven hundred acres, which he aptly called Hayward Park. There he erected a colossal Queen Anne mansion complete with towers, gables, and bays, which after his death became a popular resort hotel until its destruction by fire about 1905. Surrounded by a lake and colorful gardens, his country estate was also known for its enclosed deer park and the gas-illuminated rosewood and mahogany stable, featuring mosaic-tiled floors and a plate-glass tack room for his many carefully groomed horses. In the vicinity Hayward built an earthen dam and reservoir which supplied his estate and the town of San Mateo with pure spring water until the local Spring Valley Water Company assumed the responsibility in the late 1880s.[9]

Examples of oversized estates and garish residences were not confined to California. Other western regions contributed their share to the mansion-building mania of the mining magnates. Although many Comstock millionaires preferred to have their stately homes in distant San Francisco, at least one, Lemuel "Sandy" Bowers, constructed his magnificent memorial to a silver fortune in the Washoe Valley near Carson City, Nevada, not far from the source of his mining wealth in Virginia City. The Comstock Lode's first millionaire, Bowers erected his granite landmark in 1864. He and his wife spent over $400,000 on the project, including doorknobs and stair rods made of solid gold bullion and elaborate European furnishings and antiques.[10] When asked by a friend why he spent so much money on a house in a virtual wilderness, Bowers replied that he had "made his money in this country and... he intended to spend it here."[11]

By the turn of the century, Utah could boast as many millionaires in proportion to its population as any state in the Union. Most had made their money in mining. On East South Temple, formerly historic Brigham Street, in the most fashionable residential quarter of Salt Lake City, they constructed their mansions. These included the elegant home of Samuel Newhouse, a Bingham Canyon copper mining investor. Purchased from the Brigham Young family in 1902 for reportedly $100,000 and remodeled by its new owner, the Newhouse home contained the usual amenities of rich residents and a unique cellar of rare wines and imported liquors concealed by an electrically-operated trap door.[12] Between 1900 and 1902, Thomas Kearns built an elaborate, three-storied, marble mansion resembling a French Renaissance castle on East South Temple. Complete with thirty-two rooms in all, including a Louis XIV drawing room, a black-stained Flemish oak library, and a formal dining room of rich red-stained mahogany, this intricately decorated showpiece, which cost over $231,000 to construct and furnish, also boasted a fashionable Moorish parlor and large interior columns carved in such allegorical scenes as "The Rape of the Sabines" and Botticelli's "Graces." A social fad of turn-of-the-century America, the parlor was used exclusively for drinking after-dinner Moorish, or Turkish, coffee from elegant china cups. In contrast to the home's embellished artistic decor, a billiard room and basement bowling alley offered the simple practical pleasure of physical sport.[13] A block from Kearns's mansion, David Keith, his lifelong friend and partner in the Silver King Mining Company, erected a three-storied, limestone residence, highlighted by four massive Tuscan stone columns supporting a Greek-style portico. Interior appointments included Gothic archways, Tiffany art glass skylights, and an octagonal rotunda of polished cherry wood. A large adjacent carriage house contained servants' quarters, a bowling alley, shooting gallery, and even a pigeon coop.[14] By 1905, Matthew Walker of the Walker Brothers had completed his palatial residence on East South Temple at a cost of more than $275,000. The Spanish Renaissance home with its cream-colored stucco exterior and red Spanish tile roof was distinguished by an Aeolian organ with 1,500 gilded pipes extending from the first to the third floor. Other notable features included an expansive music room for periodic concerts, a French drawing room, a Tiffany glass skylight and chandelier, a $5,000 Otis elevator, a wine cellar, a bowling alley, and a tennis court.[15] In 1905, Enos Wall bought a two-story adobe home on East South Temple which had originally been built in 1880 by Mormon Bishop James Sharp. Within ten years, architect Richard Kletting transformed the home into "a palatial dwelling resembling a Renaissance villa."[16] Although equipped with the gilded amenities common to most mansions of the Victorian era, the Utah copper king's home was distinguished especially by its technological innovations, including the architect's use of reinforced concrete, an Otis electric elevator, a built-in vacuum cleaning system, and a steam heating plant. According to a reputable architectural journal, "The Wall Mansion, elegant and stately, has its real significance in the technology which can be discovered only on close examination—in the contrast of eclectic opulence and advanced technological features."[17]

In Butte, Montana, the legendary richest hill on earth, copper mining fortunes raised mansions in the Rockies which appeared to rival the hundred-foot-high wooden headframes above the entrance of each mine shaft. In 1888, William A. Clark built on the exclusive west side of the city a three-storied, thirty-two room Victorian-Elizabethan structure with frescoed ceilings, Tiffany windows, hand-carved woodwork and mantel pieces, and a magnificent oak staircase. The house also contained a chapel, library, billiard room and a ballroom with a pipe organ. The opulence of Clark's red brick residence contrasted sharply with the simple surroundings of the small boarding houses and framed homes of the working miners.[18] Clark's rival, Marcus Daly, preferred a home in neighboring Anaconda. There he acquired a tree-lined estate on Sixth Avenue within easy reach of his Anaconda mining operation. During his later years, Daly spent more time in far western Montana. In 1890, he built a mansion on his 20,000-acre Bitter Root stock farm. The handsome house was equipped with such attractions as a private laundry, swimming pool, greenhouse, and artificial lake.[19]

Charles A. Broadwater in later life stayed on occasion in his favorite suite at the elegant resort hotel he erected in Helena, Montana, in 1889, although it was not his private mansion or usual residence. President of both the Montana National Bank and the Montana Central branch of the Great Northern Railway as well as the owner of extensive Montana mining properties, Broadwater appropriately named his monumental creation the Broadwater Hotel and Natatorium. A contemporary Northern Pacific Railroad brochure described the hotel as "a beautiful, cottage-like structure, two stories high, with towers, balconies, verandas and wide porches."[20] The interior appointments included intricate antique oak woodwork; parlor, hall, and chamber floors carpeted in velvet; and a sumptuously furnished dining area adorned with beveled plate-glass bay windows and cut-crystal decorations. The natatorium, a great aquatic amphitheater measuring three hundred feet by one hundred feet, completed the health resort and made the therapeutic properties of hot mineral water available to the guests. This fabulous indoor "plunge," reminiscent of the enormous public bath house Adolph Sutro built in San Francisco in 1896, was reputedly the world's largest indoor swimming pool of its day. In 1892, Broadwater died in his suite at his famous health resort, only three years after its long-awaited opening.[21]

Colorado mine and smelter owners generally preferred to build their mansions in Denver rather than in the isolated mining camps where they had prospered. David H. Moffat constructed his four-storied, white brick dream house in the fashionable Capitol Hill section of the city. French tapestry–covered walls; a cathedral-like stained-glass window from Tiffany's; a drawing room with specially imported rugs, Louis XIV furniture, and a pure crystal chandelier made in Czechoslovakia; and "enough solid copper and brass in fixtures and trim to stock a foundry," highlighted the thirty-six room mansion.[22] Dennis Sheedy erected his colossal, red brick, sandstone-trimmed home on Capitol Hill. Featuring sharp Gothic gables; an expansive pillared porch; and a variety of towers, loggias, and window sizes and shapes, the Sheedy home on Grant Street has been called a "prime example of Grant Street Gothic."[23] According to Sheedy's autobiography, his home was not just another mansion on "millionaires' row," but a personal reflection of his love of nature: "I called attention in the story of my early boyhood to my love of nature, and so I have around my home sufficient land for flowers and trees, and it is with pride I have seen passers-by stop to admire the grounds. This same love of nature prompted me to furnish each room of the house with a distinctive wood, such as maple, sycamore, cherry, birch, quarter-sawed oak, mahogany, and so on, and I carried out the idea to include the furniture to correspond with the wood of each room."[24] Horace Tabor also chose to reside on Denver's Capitol Hill. In 1886, he brought his new wife, Baby Doe, to a house between Grant Street and Sherman Avenue, where they lived for the next decade. The Tabor home cost $54,000, but he spent more money turning it into one of Denver's showplaces.[25] Although local society ignored the Tabors, a contemporary observer left this flattering appraisal of their mansion: "As it stands upon this commanding eminence, in the midst of this cultivated forest, adorned by works of art, it is, as to spaciousness, embellishment, and location, the Central Park of Denver, while as an entirety it is the stateliest and most beautiful residence in this city of homes."[26] In the early 1900s, Sherman Avenue also was the site of the flamboyant house owned by the smelter king Simon Guggenheim before his election to the U.S. Senate.[27] Another U.S. Senator from Colorado, Nathaniel P. Hill, also found this Denver neighborhood a likely location for mansion building. His luxurious home with its cathedral spire and chimneys like triumphal arches was a focal point for the gathering of Denver's elite in the 1880s and 1890s. In 1908, Hill's daughter-in-law, Mrs. Crawford Hill, the undisputed leader of Denver society, attempted to give that society formal definition when she designated Denver's "Sacred Thirty-six" in her social guidebook.[28]

Some of the city's social leaders lived on Pennsylvania Street. Lined with imaginative, expensive houses of the late 1880s and 1890s, this avenue was a prominent address on Denver's Quality Hill, an eastward extension of Capitol Hill. It was here that James B. Grant, the smelting entrepreneur who became the first Democratic governor of the state (1882–86), built his mansion in 1902. Constructed high on a bluff, it combined the Colonial and Renaissance Revival elements of Victorian architecture. Designed especially for entertaining, with a combination ballroom and auditorium built for the governor's inaugural ball, the house was lavishly furnished by Mary Goodell Grant, who travelled extensively along the east coast collecting Colonial antiques for her new home. Although Grant, the son of an Alabama planter, modeled the house after southern plantation homes, the first floor contained a billiard room and bowling alley, features popular with the rich in the late nineteenth century.[29]

Not all Colorado mining leaders elected to live in the state capital. Like their California counterparts, some preferred country estates where they could escape the congestion and commotion of the city. For example, Thomas Walsh purchased Wolhurst near Denver for $150,000 in 1906. Formerly owned by the late U.S. Senator from Colorado Edward O. Wolcott, the estate contained five hundred acres, extensive gardens, a private bird preserve, stables, and a sumptuous house equipped with a library, billiard room, and art gallery. In 1909, Walsh prevailed upon President William H. Taft, whom he had politically supported in 1908 as a delegate to the Republican national convention, to visit the Colorado residence and rebaptize it Clonmel in honor of the Irish village where Walsh was born.[30]

Many more examples of the urban palaces and country estates of the bonanza kings could be cited. However, without further illustration, it is important to understand why western mining elites, like other business executives in the nation, built such big, elaborate houses. As Edward Kirkland argues, it is possible that such extravagance owed something to overly ambitious architects and socially climbing family members, and perhaps even to the responsibility of business leaders in a capitalist economy to create jobs for others.[31] Nonetheless, as Kirkland also suggests, "these palaces were erected to impress one's fellow millionaires."[32] Indeed, their concentration and high visibility in residentially segregated enclaves within a city or in scenic less-populated areas reflects the overt desire for peer-group recognition.

In addition, like the impressive opera houses, theaters, and hotels that some mining moguls built in the west, the mansion was intended in part to convince skeptical established eastern elites of the respectability of western wealth. Easterners were sometimes critical of western nouveaux riches, but westerners hoped to answer their criticism by building cities and stately mansions along the lines of those in the East and by drawing favorable comparisons between them. A contemporary Denver observer had this in mind when he wrote: "Denver is the Washington City of the west. Some of its avenues—Lincoln, Sherman, and Colfax—are more remarkable for their 'magnificent distances' than any that may be seen in Washington. . . . Many of her homes will not suffer in the least by comparison with the costliest upon famous Massachusetts Avenue."[33] A contemporary western newspaper editor reflected a similar sensitivity to eastern opinion when introducing a lengthy article on William A. Clark's Butte mansion: "In presenting the engraving and description of the newly erected residence of W. A. Clark of Butte, *The Miner* feels that it will surprise many of its readers in the east. Their idea of our city is that it is a rude mining camp, and that its structures are of the plainest and most primitive class. The reverse is the truth. Butte boasts of as handsome and as modern residences as any city in the country, and its social life will compare with that of any of the great centers."[34] For those who needed to demonstrate their wealth further or to win acceptance from eastern society, there was always the opportunity to build and live amongst the rich on the Atlantic coast.

Young men who had followed Horace Greeley's advice to go west to make their fortunes now went east as old men to publicize them. Clark exemplifies this pattern. However, the monstrously incredible mansion that he built on Fifth Avenue in New York in 1906 was so flamboyant and garish as to incite indignation and ridicule even in an age inured to such extravagance. The house, which reportedly cost $3 million, contained 131 rooms, including 21 bathrooms and 31 servants' rooms. In addition to four art galleries and a magnificent swimming pool, the residence boasted a sixteenth-century salon brought intact from Paris, the ceiling of a French chateau, and several stained-glass windows from Chartres. To construct the house Clark bought an entire stone quarry in New England.[35] Perhaps a contemporary poem by

Wallace Irwin provides the most fitting description of the Clark residence and the motives behind its creation:

Senator Copper of Tonopah Ditch
Made a clean billion in minin' and sich.
Hiked for New York, where his money he blew,
Buildin' a palace on Fift' Avenoo.
"How," says the Senator, "kin I look proudest?
Build me a house that'll holler the loudest.
None of your slab-sided, plain mossyleums!
Gimme the treasures of art an' museums!
Build it new-fangled,
Scalloped and angled,
Fine, like a weddin' cake, garnished with pills.
Gents, do your duty.
Trot out your beauty.
Gimme my money's worth—I'll pay the bills."[36]

Although offensive to many, it is uncertain how Clark's ornate stone structure affected his social ambitions in New York City. At least two biographers claim that he was unable to get into the social register.[37] Other Clark students assert that the New York social register included him as early as August 1894.[38] Whatever the case, Clark's New York residence, more than his sumptuous houses in Santa Barbara, California, and Butte, stood as a symbol of the age of excess until its destruction in 1926.[39]

California-based mining capitalists Darius O. Mills and James B. Haggin also moved to New York City after accumulating their western fortunes. In 1878, Mills retired from active business in the Far West, though retaining extensive investments there. In New York he managed his diverse business empire from a Broad Street office and occupied a spacious mansion on Fifth Avenue.[40] By the 1890s, Haggin also had opened an office on Broad Street and had a mansion near Mills on Fifth Avenue. In addition, he acquired much Manhattan real estate. The *New York Times* claimed that he "became in every sense a New Yorker."[41]

Joseph R. DeLamar was also attracted to New York in later life. In 1904, his $600,000 mansion was completed on the northeast corner of Madison Avenue and Thirty-seventh Street, opposite the home of J. P. Morgan. According to the *New York Times* the DeLamar residence was one of the costliest dwellings in the city. Not satisfied with a substantial town house, he also acquired a country estate at Glen Cove, Long Island. Toward the end of his life, he spent much time at Pembroke, his country home, where he enjoyed entertaining guests, walking in his colorful flower garden, and playing the organ. The owner of one of the finest organs in the country, DeLamar was an accomplished musician.[42]

Not all mining leaders who lived in New York City during the later stage of their careers built mansions. The Montana copper king F. Augustus Heinze made his office and residence in a double suite of luxurious rooms at the Waldorf-Astoria Hotel after liquidating his Butte properties in 1906. The fashionable Waldorf was the center of the city's social life and Heinze's suite frequently was the scene of gala parties and elaborate banquets attended by the smart set of Manhattan society. His lavish lifestyle of late hours and heavy drinking helped to exact a premature toll. In 1914, Heinze died of cirrhosis of the liver at the age of forty-four without leaving a residential monument on Fifth or Madison Avenue.[43]

Some mining elites who wanted to live in the East or move in exclusive social circles were not attracted to New York. Thomas Walsh chose eventually to reside in Washington, D.C. According to a student of the rich, the national capital "held more appeal in those days for newly solvent westerners than rock-ribbed and impersonal New York. Where the Vanderbilts and Belmonts were impervious to *arrivistes*, a senator who had been assisted into office could always be counted on to help the socially ambitious on their way up the ladder, and the Walshes were socially ambitious."[44] No doubt this explains why Walsh decided to build a magnificent mansion on Massachusetts Avenue that cost nearly $1 million. A four-storied structure with sixty rooms and twenty-three servants, the new residence included a great reception hall, ballroom, roof garden, library, promenade, galleries, and even a special apartment for the anticipated American visit of Walsh's sometime mining partner, King Leopold of Belgium. The house also was furnished with Persian rugs, Flemish tapestries, and Italian vases, revealing the tendency of American parvenus to import European culture in order to dignify their surroundings.[45] A letter from a New York importer of fine fabrics cautioned Walsh to allow only authorized personnel to unpack certain furnishings: "Mr. Bogumil is not accustomed to handling gold furniture, but is all right for work that is there now."[46] Reflecting upon the lavishly furnished mansion, Walsh's daughter, Evalyn, gave this retrospective evaluation of the social significance of the family home: "It really is a palace that expresses dreams my father and mother had when they were poor in Colorado. This should not give rise to any notion they were sadly 'nouveau,' as that term is used to label people outside of society trying to get in. We *were* [her emphasis] in; make no mistake about it."[47]

During the summer of 1905, the Walshes entered the select society of Newport, Rhode Island, where the New York upper-class vacationed amidst the gilded splendor of such palatial residences as The Breakers, built for Cornelius Vanderbilt II in 1895. The Walshes rented a suitable mansion for entertaining next to Mrs. O. H. P. Belmont's Beaulieu on prestigious Bellevue Avenue.[48] Like Walsh, James B. Haggin also was attracted to Newport society. He owned a summer "cottage," the Villa Rosa, where he died in 1914 after a month's illness.[49]

In addition to those who actually built mansions in the East or resided there later in life, there were other mining magnates who went east occasionally on business or for vacations. Still others, who served in the U.S. Senate—such as John P. Jones, Horace Tabor, James Fair, William C. Sharon, Thomas Kearns, Thomas Power, Nathaniel P. Hill, and George Hearst—established temporary living quarters on the Atlantic seaboard. Hearst even built a stately home on New Hampshire Avenue in the nation's capital.[50] Nonetheless, the prevailing view, as suggested by historian Ben Seligman, that nouveaux riches in other parts of the country were overwhelmingly drawn east like so many lemmings to live and spend their money is an exaggeration: "Most of the new rich gravitated to New York and other eastern cities, although their properties might be in Pennsylvania, Ohio, or the Far West. There was much more excitement in the big city to which these nouveau riche came as quickly as they could. There they built imitations of European mansions along the main avenues. . . ."[51] Similarly, reflecting upon a California visit, James Bryce reported in 1888 that "among the richer many look forward to a time when having made their fortunes, they will go East to spend them."[52] Although these generalizations certainly apply to some of the mining elites, the majority preferred to work and live in the West.

Indeed, some found eastern society unattractive or at least had reservations about relocating. Lloyd Tevis, one of Hearst's mining partners, was among them. In 1891, he wrote his daughter, Louise: "How curious it is that the Haggins are nearly all living in New York. . . ."[53] A year later, he warned Louise, who was then residing in New York, about the pitfalls of polite eastern society: "A certain amount of social intercourse is necessary and eminently enjoyable, but, the extravagance and excess to which it is carried within fashionable circles of New York, strikes me as destructive of both good health and rational enjoyment."[54] Montana's Marcus Daly had similar reservations about New York. After a visit he reportedly said: "It is fine for about three months. Then a person begins to get tired of their French cooking, their noise, their scramble. I find French cooking don't appeal to me. I find, too, that I do not care for the society of the leaders of the financial world. I would rather be back here talking to the lads I know in Anaconda."[55] Simeon G. Reed visited New York on business in 1884 and on other occasions for pleasure, staying at fashionable hotels on Fifth Avenue. However, like Daly, he looked forward to a return trip west rather than to a permanent residence or retirement in the East. Of one trip in 1889, he wrote his business secretary: "We [he and his wife Amanda] both feel like getting home. We have done about all the Amusements and settled down to a quiet Hotel life."[56]

Although only some bonanza kings went so far as to build a mansion in the East, the size, expense, and gilded grandeur of their western residences reflected in part a deference to eastern society. In a nation without a European-style aristocratic tradition, the older, established elites of the East often influenced the lifestyle of the newly rich elsewhere. The big, lavishly decorated and furnished house was essential for nouveau-riche mining entrepreneurs seeking to legitimize their position as wealthy, socially respectable citizens.

However, a properly equipped mansion was not in itself sufficient reason to be accepted into society in the West or the East. It had to be used as a site for the ceremonial creation of costly entertainment. Indeed, the sheer size of these cavernous castles made them inherently unlivable unless they could be filled from time to time with numerous guests. According to one student of the rich, "you couldn't build a big mansion and sit there and wait. You had to compete for notice in the new society pages of the metropolitan press."[57] As economist Kenneth Galbraith suggests, "the person seeking distinction invited his friends and competitors to his feasts, orgies or other entertainments. These were the very people he most needed to impress, those upon whose good opinion his own standing depended."[58]

Although many examples of balls, banquets, parties, and receptions could be given, a few should serve sufficiently to illustrate the prevailing pattern of entertainment among the mining rich. Local elites on the San Francisco peninsula could look forward to long and elaborate dinners at James Flood's Linden Towers. At nearby Belmont, William C. Sharon entertained former president Ulysses S. Grant in 1879 during his world tour. Grant's San Francisco stop also included a lavish luncheon of many courses at Linden Towers, where his son began a much talked about romance with Flood's daughter, Jennie.[59] Earlier the Belmont estate had been the scene of William C. Ralston's many gala parties, including his famed masquerade balls. Such visiting dignitaries as Anson Burlingame, U.S. minister to China in the 1870s, and former U.S. Secretary of State William Seward found Ralston a genial host. A young man who experienced Ralston's hospitality recalled: "He lived like a prince, and was the most bountiful entertainer I have ever known. . . . Notwithstanding all the gastronomic feats we had already accomplished, we were set down to a table loaded with viands and dainties as delicious as could have been served at Delmonico's [a fashionable Manhattan restaurant]."[60] The Ralston residence was especially well equipped for entertaining since it contained a dining room capable of seating 110 guests and a huge mirrored ballroom with a high arched ceiling and skylight. However, the latter was no match for Mrs. William Astor's more celebrated ballroom, which allegedly could accommodate four hundred guests, thus inspiring social arbiter and host Ward McAllister to invent New York's famed "four hundred" as the elite of high society.[61]

Like their California counterparts, Colorado mining magnates did not neglect their social obligations and aspirations as wealthy citizens. In 1886, in honor of her wooden (fifth) wedding anniversary, Mrs. James B. Grant gave a ball for 175 guests. A local Denver newspaper reporter commented: "Mrs. Grant is a thorough society woman, and her wooden wedding quite equalled any of her previous entertainments."[62] Judging by her diary, Nathaniel P. Hill's wife, Alice, dutifully fulfilled her role as a civic-minded hostess: "1880—First large dinner party at home—Guests—General Sherman, Clarence King... Justice Miller, [etc.]."[63] As with other rich people, a mining leader's standing in high society was in part measured by the company he kept, thus it is not surprising to find notable public figures such as General William Tecumseh Sherman and King, the first director of the United States Geological Survey, on the Hills' guest list. Ever mindful of the importance of such associations, Thomas Walsh celebrated the completion of his Washington, D.C., home with a ball in honor of President Theodore Roosevelt's spirited daughter, Alice. Musicales, receptions, dinners, and dances for the social and political leaders of Washington soon enlivened the Walsh home. By 1904, the *Denver Post* could boast that the Walshes "have demonstrated to the most skeptical that they possess the art of entertaining to the fullest degree."[64]

The same might have been said of Utah mining moguls. Although certainly not as extravagant as Walsh, Jesse Knight could also be a gracious host. According to his daughter, Jennie, his comfortable, beautiful home in Provo, Utah, served to entertain "friends and many of notable rank." The latter included William Jennings Bryan during one of his many speaking tours.[65] Apparently, Uncle Jesse, as he was affectionately known, held strong occupational and political convictions. His daughter recalls: "Father used to say, 'It's all right so long as our children don't marry school teachers, or Republicans.'"[66] It is assumed that three-time Democratic presidential candidate Bryan was entertained most hospitably. In 1903, another politician, President Theodore Roosevelt, was handsomely honored at the Salt Lake City home of Thomas Kearns. Luncheon was served in the formal dining room, which was appropriately decorated with American Beauty roses and Roosevelt carnations. The broad-minded but naive Kearns invited a Catholic bishop, an Episcopal bishop, and a Mormon apostle and his plural wife to meet the president. The bemused Roosevelt later observed that the conversation included "every combination of beliefs and systems of thought and civilizations that were ages apart."[67]

Entertainment did not have to be confined to the big house. Private, personalized railroad cars, which were popular with the rich, could be used for social engagements. Samuel Newhouse, who had a reputation as one of the best party-givers on the railroad tracks in the country, took this practice to extremes when he reportedly hired an entire private train to give a party. According to his biographer, "this was on the Western Pacific Railroad. The party went from Salt Lake to either Reno or Sacramento and back. The party and the money spent was unbelievable. The food and drinks were lavish, and he invited his friends from all over the world to join him."[68]

Newhouse's imagination as a host was perhaps matched by Joseph R. DeLamar's expenditures. In 1892, he gave a magnificent dinner on the East Coast in honor of President Benjamin Harrison's son, Russell. The dinner was followed by a reception and ball attended by nearly two hundred guests. DeLamar became widely known for his dinner parties. According to some contemporaries, "he generally had filled fountain pens as favors in front of every plate so a pen would be handy for anyone wishing to purchase stock in DeLamar enterprises."[69] If true, such obviously self-serving tactics were atypical. Social engagements were undoubtedly good for business, but most mining leaders saw them as an opportunity to display their wealth rather than to increase it.

The acquisition and exhibition of an art collection was another socially acceptable way of using one's mansion for the display of wealth, especially because business elites of the era generally collected works of art that were distinguished more by their cost than their artistic merit. It has been suggested that "rich Americans spent more money on art during the thirty years from 1880 to 1910 than had ever been spent by a similar group in the world's history."[70] In the process, the unsophisticated American upper-class was often victimized by art dealers who sold spurious canvases as Old Masters. The collections of J. P. Morgan and William Vanderbilt revealed their lack of artistic appreciation. Although he fancied himself a connoisseur of art, Vanderbilt's whole collection contained only one reputable painting, a Delacroix.[71] According to a student of the rich, "he knew nothing of art, and underneath his pretensions cared less, for he was a complete utilitarian; but it had become fashionable to have an elaborate art gallery, and he forthwith disbursed money right and left to assemble an aggregation of paintings. . . . Not a few of the paintings unloaded upon him were, despite their high-sounding reputations, essentially commonplace subjects, and flashy and hackneyed in execution; but he gloried in the celebrity that came from the high prices he was decoyed into paying for them."[72] However poorly informed, the purchase of paintings by eastern business leaders apparently gave impetus to similar acquisitions by newly rich western mining entrepreneurs.

Upon completion of his memorable mansion, Thomas Walsh was approached by an art dealer who hoped to interest him in purchasing $20,000 worth of alleged masterpieces: "They are all masterpieces and I know that if you were to see them you would buy them. The original owner at one time was a multi-millionaire."[73] Since the canvases were not identified by their respective titles or painters but with a wealthy owner, one may surmise that affluent individuals were likely to purchase art because others of similar status had done so. For such people it was more important to appear cultured than to develop a truly discriminating taste for the so-called finer things in life.

This may explain why naive mining investors could be easily deceived by art dealers. Simeon G. Reed proudly displayed twenty-four canvases from his private collection at the Portland

Mechanic's Fair in 1883. However, his biographer asserts that the "most remarkable thing about the collection was the record of the prices paid for these mediocre works."[74] Albert K. P. Harmon was similarly not very astute in his early art acquisitions. Reportedly, "he was at first so unskilled in judging the values of pictures that he accumulated a deal of trash, and when he one day asked a candid friend for an opinion of his pictures, the latter jocosely told him their best use would be to stop holes in the roof."[75] According to a contemporary source, Matthew Walker of Utah owned a collection containing the works of local and foreign artists, "among them paintings by the following eminent French masters: Barillot, Perrault, Japy, Ravanne, Petitjean, Charpin, Chabas, and Chretein."[76] Assuming they ever existed, these individuals could not be classified as master painters. Therefore, it is very possible that Walker also was a victim of unscrupulous art merchants who preyed upon the rich.

Other mining entrepreneurs gained reputations as art collectors. Joseph R. DeLamar was said to have "assembled a notable collection of art works."[77] Henry R. Wolcott of Colorado was described as a "litterateur, art lover."[78] Although he may have possessed an art collection, it would be an exaggeration to describe him as a man of letters. Thomas Selby juxtaposed interior and exterior aesthetic pleasures in his expansive country estate, Fair Oaks, in California's San Mateo County. Long French windows separated art galleries on the inside from flower-filled gardens outside the house.[79]

A number of mining elites, like other rich business executives who sought to patronize the arts, found Europe a favorite hunting ground for their avocation. During his later years, Darius O. Mills spent much time in foreign travel, collecting works by Gérôme, Detaille, and Dauguereau which were displayed at his Millbrae estate.[80] William A. Clark, who owned a Paris apartment, often went to Europe on art collecting excursions. He even served as vice-president of the French Institute of the United

States, an organization to encourage interest in French art through exhibits of French paintings, laces, ceramics, and decorative arts.[81] Adolph Sutro, who is better known for his private library than his art collection, nonetheless spent considerable time abroad seeking the artistic treasures of the Old World. He had copies made of classical and modern European statues of mythological creatures, animals, gnomes, elves, and famous historical figures for display at Sutro Heights, his beautifully landscaped, San Francisco garden estate overlooking the Pacific Ocean. On one trip abroad he was so taken with an enormous statue entitled *Triumph of Light* by the Belgian artist Wirtz that he purchased the piece and had it sent in sections to San Francisco. He renamed it *Liberty Enlightening the World* and conspicuously placed it in 1887 on a fourteen-foot-high pedestal at the top of Mount Olympus near the center of San Francisco. Reportedly, Sutro was interested in exposing the bohemian city by the bay to the instructional value of his various statues.[82]

Although there was a prevailing tendency among the mining moguls to buy paintings and objets d'art that were both costly and of European origin, at least one mining investor tried to sustain culture on native grounds. William C. Ralston commissioned a local artist, G. J. Denny, to decorate the interior of the California Theatre he built in San Francisco. Denny filled the paneling between the theater boxes and the front of the curving balconies with colorful paintings. He also executed a monumental painting of the bay from the vantage point of the Golden Gate Bridge on the drop curtain of the stage.[83]

An interest in art occasionally was pursued more ardently by mining wives than by their husbands. According to a notation in her diary, Mrs. Nathaniel P. Hill read her article on Peter Paul Rubens, the greatest Flemish painter of the seventeenth century, at a Denver Art Club meeting in 1879.[84] Marie Mackay reputedly was generous to struggling artists.[85] This trait she shared with Phoebe Hearst, who not only subsidized the education of art students abroad, but assembled an impressive art collection of her own. Shortly after her marriage to George, their mansion on Van Ness Avenue in San Francisco was equipped with a gallery containing her European art and statuary treasures. The family home also became a site for the entertainment of painters, musicians, and writers. While George served in the U.S. Senate in the late 1880s, Phoebe added an art gallery to their Washington mansion, which was decorated with rare paintings, tapestries, and artistic objects she collected in Europe. In 1916, she proudly showed her art collection at the Palace of Fine Arts in San Francisco, which had been designed by her friend the famous architect Bernard Maybeck for the Panama-Pacific International Exposition of 1915.[86] It should not be assumed, however, that the mining wives predominantly were art connoisseurs or collectors. Most occupied their time looking after their children, hosting or attending society affairs, and promoting various acts of charity.

Although a few mining leaders, such as William A. Clark, took a serious, albeit very expensive, interest in art, and such others as William C. Ralston and Darius O. Mills supported art associations and museums, most mining elites apparently regarded the art collection as a socially obligatory decoration for their private mansions. The more they could afford for paintings, statues, and antiques to fill their galleries the better, since the very wealthy of the Gilded Age often spent money for the sake of proving to others like themselves that they had money to spend.

Virtually all Victorian mansions worthy of the name had a room set aside as a library, and many of the mining moguls acquired extensive book collections. Considering the donations of some to educational institutions, including public libraries, it is reasonable to assume that their private libraries reflected at least in part a sincere personal interest in academic knowledge. For such individuals the home library was meant to be read rather than simply to be treated like an expensive art collection.

Winfield S. Stratton discovered the cultural enrichment of leisure reading in his Colorado Springs home. According to his biographer, "the library became filled with books. He grew particularly fond of Balzac. Then he began reading theology, religion and metaphysics. The shelves were lined with his favorite works: Pascal, Henry Drummond, Bible commentaries, the writings of the Duke of Argyll, and Fiske's *The Destiny of Man*, *Through Nature to God*, and *The Idea of God*."[87]

Other Rocky Mountain mining entrepreneurs apparently shared Stratton's fascination for books. When David H. Moffat came to Denver in 1860, he opened a book and stationery store with C. C. Woolworth. Although Moffat's formal education did not extend beyond the eighth grade, his Denver mansion included "a great library lined with books, many from his own bookstore." However, he never found sufficient time to read his extensive collection.[88] David Keith of Utah also received a limited education, but he allegedly became familiar with his books. A contemporary source claimed that "he finds a recreation in reading, his library of standard works being one of the most complete in the West."[89] Quiet and mild of manner, Keith was apparently "more comfortable chatting with a few close friends in his library than mingling with many guests."[90] Colorado's Irving Howbert was also known for his book collection. A contemporary biographical reference reported that his "private library is one of the finest in the State, and he reads the books, so that his general information covers a wide range of literary and scientific subjects as well as history."[91]

Among the mining elites, no one surpassed Adolph Sutro's energy and enthusiasm for book collecting, a passion that he began as a boy in his native Prussia. A true bibliophile, he had accumulated by 1895 about 250,000 works for what was reputed to be the largest private collection in the United States. Many of the volumes he acquired abroad during his extensive European travels in the 1880s. The library contained much scientific and technical literature, about 3,000 incunabula produced between 1455 and 1500, and many manuscripts, including the private papers of Sir Joseph Banks, the eighteenth-century British botanist who circumnavigated the world with James Cook. Sutro hoped eventually to create the Sutro Library on his San Francisco property to house the collection, thus making it available for research by visiting scholars.[92] His passion embraced books for the young reader as well as those for the serious scholar. He once said that "good books are the best and safest companions to be selected by young people."[93] About half of his collection was destroyed in the San Francisco earthquake and fire of 1906. Only forty-two incunabula, the Banks papers, 90,000 volumes, and a large pamphlet collection survived. They were donated to the state of California by Sutro's heirs in 1913 and became the core of a branch of the state library in San Francisco named after Sutro, thus posthumously fulfilling his earlier plans for a research facility.[94]

Although Stratton, Moffat, Keith, Howbert, Sutro, and others developed libraries for their own personal pleasure and education, most mining entrepreneurs probably bought books because it was fashionable for the wealthy of the era to have a well-stocked library. This most likely explains the motivation, for example, of such mining leaders as Thomas Walsh of Colorado, James Flood and William C. Ralston of California, Sandy Bowers and Simeon Wenban of Nevada, and Enos Wall of Utah. With the possible exception of the latter, none of these men was known as an avid reader and at least one could not read at all, yet their mansions all contained extensive libraries.

In 1903, Walsh received an inquiry from a New England publishing company that stocked private libraries with the "various editions and bindings of the standard authors."[95] It is not known if he purchased any books from this dealer, but the emphasis in their libraries on standard authors or the classics suggests that potential buyers among the upper class were more interested in creating a positive social impression with their book selections than in reflecting their own individual literary tastes, if they had any. This may explain why Ralston's library in his Belmont home contained "leatherbound books chosen for their looks rather than their content."[96]

The library of Bowers especially testifies to the social significance of book collecting. He and his wife ordered morocco-bound books, with their titles imprinted in gilt letters, by the dozens of sets, for their Nevada mansion. The purchase was particularly ironic. An illiterate former teamster, Bowers was unable to read the fine volumes that stocked his shelves, but he nonetheless had created the image of a cultured gentleman, thereby fulfilling the contemporary social expectation that a man of wealth should also be a man of letters.[97] In addition, book buying allowed western mining entrepreneurs to join the select company of eastern business leaders with a strong commitment to libraries, such as J. P. Morgan and Andrew Carnegie. Like the art collection, the library was an essential complement to the mansions of America's industrial rich.

When the brilliant English courtroom lawyer and judge Sir Edward Coke asserted in the early seventeenth century that a man's house is his castle, he obviously was not referring to the mansions of the Gilded Age. Yet, his statement literally describes the palatial houses of western mining elites in America's Victorian era. Of course, not every mining mogul found it necessary to proclaim his success by building a huge home. Jerome Chaffee, David H. Moffat's mining partner, once remarked, "Let other men build houses; I will put my money into mines."[98] However, with the exception of a few such dissenters, the mining leaders as a group expressed an enthusiasm for mansion-building. Although more committed to permanent residence in the West than commonly believed, the mining businessmen drew much of their inspiration for the construction, utilization, and decoration of their mansions in the West or the East from established eastern society. In an age especially enamored of the exhibition of great wealth, the big house with its manicured gardens, elegant drawing room, majestic ballroom, imported furnishings, expensive art collection, and voluminous library was the ultimate status symbol. For the nouveaux riches in particular, their imposing urban edifices and country "cottages" clearly demonstrated to themselves and hopefully to others whom they were trying to impress that they were solidly and properly prosperous and thus socially acceptable.

Although built essentially for show, the homes of the mining entrepreneurs often possessed state-of-the-art technology in lighting, water service, heating, and other electrical or mechanical conveniences. According to an architectural historian of the mining West, "the greatest contribution of the Victorian age to residential architecture was to supply ample light, heat, and water to the home. Victorian houses were the first to feature central heating by warm-air furnaces, hot and cold running water, cooking ranges, and indoor toilets, all of which were in use by the 1870s."[99] At the same time, bowling alleys, billiard rooms, tennis courts, and swimming pools—amidst the almost ethereal elegance of ceiling frescoes, crystal chandeliers, and cathedral-like stained-glass windows—invited residents and guests to enter a more mundane recreational world. In short, the stately homes addressed the practical and physical needs of their inhabitants, while the elaborately ornate decor presumably offered a spiritual or aesthetic adventure in living. Such was the paradox of the good life for those caught up in the mansion-building mania of the Gilded Age.

NOTES

1. Carnegie, "Wealth," 653, 661.

2. For a revealing photograph contrasting the humble weaver's cottage of Carnegie's birth with Skibo Castle, his home later in life, see Harold C. Livesay, *Andrew Carnegie and the Rise of Big Business* (Boston: Little, Brown and Co., 1975), 2.

3. Harold Kirker, *California's Architectural Frontier: Style and Tradition in the Nineteenth Century* (Salt Lake City: Peregrine Smith, Inc., 1973), 92.

4. *San Francisco Chronicle*, 27 March 1966; Charles Lockwood, *Suddenly San Francisco: The Early Years of an Instant City* (San Francisco: The San Francisco Examiner, 1978), 139–40; William N. Abeloe et al., *Historic Spots in California*, 3d ed. (Stanford: Stanford University Press, 1966), 364; and S. Allen Chambers, ed., *Discovering Historic America: California and the West* (New York: E. P. Dutton and Co., 1982), 110. The digging of the unsightly Second Street "cut" in 1869 signaled the beginning of Rincon Hill's demise as San Francisco's most fashionable neighborhood. For a well-illustrated, brief description of the physical and economic transformation of this district, see Charles Lockwood, "Rincon Hill Was San Francisco's Most Genteel Neighborhood," *California History* 58 (Spring 1979): 49–61. See also Albert Shumate, *A Visit to Rincon Hill and South Park* (San Francisco: Yerba Buena Chapter, E. Clampus Vitus, 1963).

5. Lewis, *Silver Kings*, 227–28; Brechin, "A Little Place in the Country," 12–14; and the *San Francisco Call*, 11 January 1880.

6. Clark E. Carr, *My Day and Generation* (Chicago: A. C. McClurg and Co., 1908), 53. See also Lavender, *Nothing Seemed Impossible*, 251–56.

7. Abeloe et al., *Historic Spots in California*, 406–7; and Daphne Reece, *Historic Houses of California* (San Francisco: Chronicle Books, 1983), 28–29.

8. Abeloe et al., *Historic Spots in California*, 402–3; Brechin, "A Little Place in the Country," 14; and *San Francisco Chronicle*, 6 August 1970.

9. Davenport Bromfield, "Alvinza Hayward, Pioneer," n.d., San Mateo County Historical Association, San Mateo, California; *Alta California*, 15 January 1871; *San Francisco Chronicle*, 20 October 1970; Abeloe et al., *Historic Spots in California*, 404; and Brechin, "A Little Place in the Country," 14.

10. *Territorial Enterprise* (Virginia City, Nevada), 6 May 1892; and Angel, *History of Nevada*, 622.

11. "Sandy Bowers," in "California Biography," n.d., unnumbered pages, Bancroft Library. For popular accounts of the mansion and Eilley O. Bowers, Sandy's wife, see Swift Paine, *Eilley Orrum, Queen of the Comstock* (Palo Alto: Pacific Books, 1949); Alice B. Addenbrooke, *The Mistress of the Mansion* (Palo Alto: Pacific Books, 1950); and Marcia Gille, "Lady Luck and the Peepstone Palace," *True West* 31 (January 1985): 42–46.

12. Margaret D. Lester, *Brigham Street* (Salt Lake City: Utah State Historical Society, 1979), 57; and Rudd, "Samuel Newhouse," 303.

13. Lester, *Brigham Street*, 91–92; and Larsen, "The Life of Thomas Kearns," 151, 180–83.

14. Lester, *Brigham Street*, 80–81; and *Biographical Record of Salt Lake City and Vicinity*, 587.

15. Lester, *Brigham Street*, 190–92; Whitney, *History of Utah*, 4:626–27; and Bliss, *Merchants and Miners in Utah*, 271.

16. Lester, *Brigham Street*, 44–45.

17. Donald J. Bergsma, "Surprising Structure behind a Classic Facade," *Western Architect and Engineer* 219 (May 1970): 32–33. See also Lester, *Brigham Street*, 45.

18. A detailed report on the Clark mansion was published in the *Holiday Miner* (Butte), 1 January 1888. For a brief description of the mansion as it exists today, see "The Gilded Age of Butte, Still in Evidence," *Sunset* 171 (July 1983): 30–34. A very useful overview of Butte society is available in Malone, *The Battle for Butte*, 57–79.

19. Malone, *The Battle for Butte*, 81; and H. Minar Shoebotham, *Anaconda, The Life of Marcus Daly, the Copper King* (Harrisburg: Stackpole Co., 1956), 149.

20. As quoted in Nedra Bayne, "The Broadwater: Relic of Elegance," *Montana: The Magazine of Western History* 19 (Summer 1969): 62. For brief overviews of Broadwater's career, see the *New York Times*, 25 May 1892, and the *National Cyclopedia of American Biography*, 45:174–75.

21. Bayne, "The Broadwater: Relic of Elegance," 61–63.

22. Edith E. Kohl, *Denver's Historic Mansions: Citadels to the Empire Builders* (Denver: Sage Books, 1957), 63, 69–70.

23. Sandra Dallas, *Cherry Creek Gothic: Victorian Architecture in Denver* (Norman: University of Oklahoma Press, 1971), 56. See also Kohl, *Denver's Historic Mansions*, 61.

24. Sheedy, *Autobiography*, 59.

25. Smith, *Horace Tabor*, 254; and Bancroft, *Chronicles of the Builders of the Commonwealth*, 4:314.

26. Henry D. Teetor, "Honorable H. A. W. Tabor, Ex–United States Senator, and Ex–Lieutenant-Governor of Colorado," *Magazine of Western History* 9 (January 1889): 272.

27. Edwin P. Hoyt, Jr., *The Guggenheims and the American Dream* (New York: Funk and Wagnalls, 1967), 182, 187.

28. Dawson, Scrapbooks, Public Men, I, 61:181; W. H. Kistler Stationery Co., *Who's Who in Denver Society, 1908* (Denver: W. H. Kistler Stationery Co., 1908); and Dallas, *Cherry Creek Gothic*, 41.

29. Kohl, *Denver's Historic Mansions*, 211, 215; Dallas, *Cherry Creek Gothic*, 59; and Eve Marie Utne, ed., *Victorian Traveler's Companion* (Philadelphia: The Victorian Society in America, 1980), 19.

30. McLean, *Father Struck It Rich*, 131, 141, 159–60; Dawson, *Life and Character of Edward Oliver Wolcott*, 1:586–90; and the *Denver Post*, 25 February 1932. Clonmel is the county seat of Tipperary, Ireland.

31. Kirkland, *Dream and Thought in the Business Community*, 33–38.

32. Ibid., 41.

33. Teetor, "Honorable H. A. W. Tabor," 272.

34. *Holiday Miner*, 1 January 1888.

35. Robert Littell, "The House that Clark Built," *New Republic* 30 (30 March 1927): 171–72; *New York Times*, 3 March 1925; and Mangam, *The Clarks*, 83–85.

36. Wallace Irwin, "Senator Copper's House" (ca. 1911), in *New York Times Book Review*, 1 January 1939, 15.

37. High, "William Andrews Clark, Westerner," 216; and Farrell, "William Andrews Clark," 72.

38. Richard Ruetten, "In Re: William Andrews Clark," *Arizona and the West* 3 (Winter 1961): 170; and Mangam, *The Clarks*, 87.

39. Littell, "The House the Clark Built," 171; *New York Herald Tribune*, 3 March 1925; and Mangam, *The Clarks*, 83, 88.

40. Johnson et al., eds., *Dictionary of American Biography*, 13:7; and *Who's Who in America*, 5:1308.

41. *New York Times*, 13 September 1914; and Samuel E. Moffett, "James Ben Ali Haggin," *The Cosmopolitan* 33 (May–October 1902): 166–67.

42. *New York Times*, 2 December 1918.

43. Ibid., 5 November 1914; Malone, *The Battle for Butte*, 191; and Sarah Ann McNelis, "The Life of F. Augustus Heinze" (M.A. thesis, Montana State University, 1947), 186–96.

44. Beebe, *The Big Spenders*, 350.

45. McLean, *Father Struck It Rich*, 91–94, 110. See various letters to Walsh from merchants and forwarding agents, Business Correspondence, 1887–1903, container 77, file 1902, Walsh Papers.

46. Letter, Neuman and Co. to Walsh, 5 October 1903, file 1093, ibid.

47. McLean, *Father Struck It Rich*, 91.

48. Ibid., 122–26; and Beebe, *The Big Spenders*, 357.

49. Johnson et al., eds., *Dictionary of American Biography*, 8:84; and *New York Times*, 3 September 1914.

50. Older, *George Hearst*, 210.

51. Ben B. Seligman, *The Potentates: Business and Businessmen in American History* (New York: Dial Press, 1971), 209.

52. James Bryce, *The American Commonwealth*, 3d ed., vol. 2 (London: Macmillan and Co., 1893), 446.

53. Lloyd Tevis to Louise Teris, 16 March 1891, Part II, box 5, William Sharon Family Papers, Bancroft Library, University of California, Berkeley.

54. Lloyd Tevis to Louise Teris, 25 December 1892, ibid.

55. As quoted in Shoebotham, *Anaconda, The Life of Marcus Daly*, 176.

56. Reed Papers, 28: 26. See also 16:93.

57. Stewart H. Holbrook, *The Age of the Moguls* (Garden City, N.Y.: Doubleday and Co., 1954), 330.

58. Galbraith, *The Age of Uncertainty*, 65.

59. Lewis, *Silver Kings*, 229–31. Jennie Flood's real name was Cora Jane.

60. Carr, *My Day and Generation*, 49, 53.

61. Lavender, *Nothing Seemed Impossible*, 255–57, 280. For McAllister's evaluation of high society, see Samuel Ward McAllister, *Society as I Have Found It* (New York: Cassell, 1890).

62. *Tribune-Republican* (Denver), 24 January 1886. For an example of previous entertainments, see *Rocky Mountain News*, 1 July 1883.

63. Hill Papers, folio 6, notebook 5, p. 23. For other parties and receptions hosted by the Hills, see *Rocky Mountain News*, 28 August, 20 November 1881.

64. See a clipping from the *Denver Post* (1904), in Dawson, Scrapbooks, Mining II, 23:265.

65. See manuscript biography of Amanda McEwan Knight by her daughter, Jennie K. Mangum, box 168, Knight Papers, 13–14.

66. Ibid., 13–14.

67. As quoted in Lester, *Brigham Street*, 94–96.

68. As quoted from an interview with J. Holman Waters in Rudd, "Samuel Newhouse," 303–4.

69. See Barrett, "When Big Money Came to Owyhee," 26–27, and note 105.

70. Cochran and Miller, *The Age of Enterprise*, 257.

71. Josephson, *The Robber Barons*, 342–43; Seligman, *The Potentates*, 211. For a discussion of art dealers, see Kate Simon, *Fifth Avenue: A Very Social History* (New York: Harcourt, Brace and Jovanovich, 1978), 226–42; and Germain Seligman, *Merchants of Art, 1880–1960* (New York: Appleton-Century-Crofts, 1961).

72. Gustavus Myers, *History of the Great American Fortunes*, vol. 2 (Chicago: Charles H. Kerr and Co., 1910), 2:213.

73. Unsigned letter to Walsh, 23 November 1902, Business correspondence, 1887–1903, container 77, file 1903, Walsh Papers.

74. Lloyd Gouvy Lyman, "Simeon Gannett Reed, A Preliminary Biographical Study" (B.A. thesis, Reed College, 1948), 64. The list of paintings is available in Reed Papers, 16:65–66.

75. See the *Oakland Enquirer*, 30 April 1896.

76. Whitney, *History of Utah*, 4:627.

77. *National Cyclopedia of American Biography*, 28:257–58. See also *New York Times*, 2 December 1918; and Hall, *America's Successful Men of Affairs*, 1:178.

78. Dawson, Scrapbooks, Public Men, 61:333.

79. Tilton, *William Chapman Ralston*, 192; Hubert H. Bancroft, "Biographical Sketches of Thomas Henry Selby," ca. 1890, Bancroft Library, 5.

80. Johnson et al., eds., *Dictionary of American Biography*, 13:7; Tilton, *William Chapman Ralston*, 194; and Hubert H. Bancroft, "Biography of Darius O. Mills," ca. 1885, Bancroft Library, 5.

81. Mangam, *The Clarks*, 54–55, 82–87.

82. Stewart, *Adolph Sutro*, 180–81, 187; and James Delgado, "Sutro Heights Park: Traces of a Dream," *American West* 18 (May/June 1981): 63.

83. Lavender, *Nothing Seemed Impossible*, 261.

84. Hill Papers, folio 6, notebook 5, p. 21.

85. Manter, *Rocket of the Comstock*, 207.

86. Richard H. Peterson, "Philanthropic Phoebe: The Educational Charity of Phoebe Apperson Hearst," *California History* 64 (Fall 1985): 284, 286; and Older, *George Hearst*, 210.

87. Waters, *Midas of the Rockies*, 188.

88. Kohl, *Denver's Historic Mansions*, 63, 70; Johnson et al., eds., *Dictionary of American Biography*, 13:75; Glenn C. Quiett, *They Built the West: An Epic of Rails and Cities* (New York: Cooper Square Publishers, 1965), 162–63; and Mehls, "David H. Moffat, Jr.," 18–19.

89. Otheman Stevens, ed., *Press Reference Library, Notables of the Southwest* (Los Angeles: Los Angeles Examiner, 1912), 111.

90. Lester, *Brigham Street*, 86.

91. Byers, *Encyclopedia of Biography of Colorado*, 207.

92. Stewart, *Adolph Sutro*, 11, 178–79, 187–88; Hart, *A Companion to California*, 435; Johnson et al., eds., *Dictionary of American Biography*, 18:224; and Delgado, "Sutro Heights Park," 63.

93. *Alta California*, 1 January 1880.

94. Hart, *A Companion to California*, 435; and Historical Records Survey, *Guide to Depositories of Manuscript Collections in the United States: California* (Los Angeles: Southern California Historical Records Survey Project, 1941), 38. For an overview of the Sutro Library's holdings, see Richard H. Dillon, *The Anatomy of a Library* (Sacramento: California State Printing Office, 1957).

95. Mrs. Paige to Walsh, 9 December 1903, Business correspondence, 1887–1903, container 77, file 1903, Walsh Papers.

96. Lavender, *Nothing Seemed Impossible*, 255.

97. Addenbrooke, *The Mistress of the Mansion*, 18–19.

98. As quoted in Dawson, Scrapbooks, Public Men, I, 61:4.

99. C. Eric Stoehr, *Bonanza Victorian: Architecture and Society in Colorado Mining Towns* (Albuquerque: University of New Mexico Press, 1975), 36.

4

The Pleasures and Diversions of the Rich

ALTHOUGH THE MANSION WAS FOR MANY of the mining elites the focal point of an extravagant lifestyle consisting of masquerade balls, dinner parties, and the exhibition of art and book collections, the pleasures of the rich were pursued just as passionately away from the home. The status symbols and diversions of the Gilded Age wealthy included the raising and racing of thoroughbred horses, the purchase of ocean-going yachts and expensive automobiles, membership in exclusive gentlemen's clubs, a seat in the U.S. Senate, the grand tour of the Old World, the occasional search for a titled son-in-law, prestigious education for the children, and the financial indulgence of family members. Western mining moguls and their business counterparts on the East Coast largely were attracted to the same customary expenditures of time and money because the proof of a business executive's social respectability was not so much how or where he made his fortune but how he spent it. In fact, according to the contemporary journalist E. L. Godkin, the rich person "has been held bound to spend his money in ways in which the public in general expected him to spend it, and in which it had become usual for men of his kind to expend it. His expenditure was, therefore, in a certain sense, the product of the popular manners."[1]

Among various expensive diversions, organized sports attracted the interest of those with money. Thorstein Veblen noted "the facility with which any new accessions to the leisure class take to sports, and hence the rapid growth of sports and of the sporting sentiment in any industrial community where wealth has accumulated sufficiently to exempt a considerable part of the population from work."[2] According to a recent student of the subject, "using their sports as badges of social status, the ultra-rich generally confined themselves to pursuits whose cost put them out of reach of ordinary Americans."[3] Horse breeding and racing was such a sport. In the United States, the late nineteenth and early twentieth centuries witnessed important developments in this pastime, including the opening of new tracks, the organization of jockey clubs in major cities, an increase in the number and purses of various races, and the emergence of what today is known as the Triple Crown. The Belmont Stakes was first held in 1867, followed by the Preakness in 1873, and the celebrated Kentucky Derby in 1875. In 1894, American racing leaders formed the Jockey Club to regulate the sport and supervise the registration of thoroughbreds. By then, many eastern and western millionaire horse breeders, including some bonanza kings, had invested heavily in what one historian calls the "commercialization of the turf."[4]

On his Bitter Root Farm in Montana Marcus Daly raised trotting and running horses. The former included such reputable brood mares as Rachel, Ellorce, and Red Silk. However, Daly's real passion was thoroughbred racing horses. Of these Tammany was the most famous and his owner's favorite. A long-shot winner in his first race, the Eclipse Stakes, as a three-year-old this fine horse won all his stakes races and as a four-year-old defeated the famous eastern horse Lamplighter. Other notable Daly horses included Ogden, which captured the 1896 Futurity at the Sheepshead Bay course on Long Island, and Hamburg, which won $75,000 in 1898. The Daly stables did not produce a Kentucky Derby winner, but Ogden was the grandsire of Zev, which won the Derby in 1923. As the Daly horses won national recognition they became a source of pride for the mining population of Butte, which could share with their prominent local citizen a day at the large race tracks built in Butte and Anaconda.[5]

Judging by the profits he earned from the racing and sale of his horses, Daly obviously took the sport seriously. His involvement even included the presidency of the Montana Jockey Club. In 1901 his estate auctioned off 369 horses for nearly three-quarters of a million dollars. This was "the greatest racing stock sale America had ever seen."[6] Nonetheless, it would be an exaggeration to assume that Daly regarded horse breeding and racing as essentially a business investment. On the contrary, he found the activity a pleasant and welcome diversion from the cares of business: "The awful work thrown upon me by my business would kill me but for the fact that now and then I can run up to the Bitter Root and see the colts. That place has cost me a lot of money, but it is worth it to me when I stop to think of the relief it furnishes."[7]

James B. Haggin, Daly's associate in the Anaconda and other mines, was perhaps even more famous in racing circles. He maintained a 44,000-acre horse-breeding farm, Rancho del Paso, near Sacramento. In the 1880s, his thoroughbreds captured many of the leading races in the country, including the Monmouth, Suburban, American Derby, and the Junior Championship. One horse aptly named Ben Ali was good enough to win the Kentucky Derby. In Kentucky, Haggin owned extensive racing stables and breeding farms. At one time he reportedly owned three times as many thoroughbreds as anyone else in the world.[8]

Like his business associates Daly and Haggin, George

Hearst was a racing devotee. He owned a five-hundred-acre horse ranch near the race track at Pleasanton, California, east of San Francisco Bay. Although his philanthropic wife, Phoebe, regarded the sport as a vulgar waste of money, her husband accumulated a stable that included such winning horses as Gorgo, Ballarat, Yosemite, and Tournament. Others, such as King Thomas, for which he paid the then substantial sum of $50,000, were disappointing. Overall, Hearst's racing career was not a profitable pastime. He probably lost more than a million dollars on his racing stable. However, according to his biographers, like a game sportsman, he did not complain.[9]

Simeon G. Reed also devoted much time and money to horse breeding and racing. He owned an extensive stable of trotting horses, including a prize mare, Jane L, which created quite a sensation in the Pacific Northwest during the trotting season of 1887.[10] During the 1892 season he was especially impressed by the performance of a stallion he had named Coeur d'Alene after the district where his Bunker Hill and Sullivan Mine was located. In a letter to his business secretary and nephew, Martin Winch, he remarked: "Yesterday went out to Pleasanton. A track outside of San Francisco. Hickok drove Coeur d'Alene a mile in 2-28, the first half in 1-20 and the last half in 1-08. He is a great Horse and everybody admires him and his gait."[11] Reed's venture into raising and racing thoroughbred running horses was less fulfilling. Toward the end of his life, he lamented: "My experience thus far has been that Horse Racing is a very expensive luxury without much fun in it either for me."[12]

Although not as notable as Hearst, Haggin, Daly, and Reed, other mining leaders were interested in horse racing. Alvinza Hayward, who raised thoroughbreds, and James Flood had private race tracks on their California country estates. While William Ralston apparently was not attracted to formal horse racing, he used his Belmont property to indulge his passion for raising high-blooded horses that could trot for sustained periods at a rate of 12.9 to 16 miles per hour while pulling a light carriage. On more than one occasion he delighted in "racing" the train over the twenty-two miles from San Francisco to Belmont.[13] Although Colorado's Winfield S. Stratton did not own a stable of race horses, he allowed himself to be selected president of the racing association created by fellow Cripple Creek mining millionaires. The organization was responsible for building the race track, grandstand, and clubhouse in Colorado Springs, a favorite resort community for Colorado's elite.[14]

While horse racing was essentially a private affair, it occasionally became a matter of public controversy. In the 1880s, mining entrepreneurs Adolph Sutro, Alvinza Hayward, and James Fair joined railroad barons Leland Stanford and Charles Crocker and others in organizing a special interest group to pressure the commissioners of the new Golden Gate Park in San Francisco into authorizing a speed track for the show and competition of their race horses. The park's designer and first superintendent, William Hammond Hall, objected to the plan, but the influential horse owners eventually prevailed after agreeing to pay part of the cost of construction. Those who regarded the race track as a violation of the true public purpose of the park as recreation for the city's ordinary residents made the case for public over private interest in this biting editorial:

> The gentlemen who are able to possess fast horses do not stand in need of this refreshment as much as some of their less-favored fellows. Their winters are passed in the sunshine of the South and their summers in villas at Lennox, or cottages by the sea. But to the poor and the children of the poor, the park offers the only glimpse of greensward that greets their eyes from one year's end to the next. It seems a cruelty to destroy these pictures of peace that a wise forethought has prepared for them especially to enable a few opulent citizens to enjoy their cho-

> sen pastime. . . . Some of these gentlemen have famous picture galleries, and all right-minded persons would sympathize with their horror and distress if some vandal hand should cut out a strip of the border of one of their favorite landscapes. But the living picture is just as truly a work of art as the painted, and the cutting away of this broad stretch of verdure and substituting for it something entirely incongruous with its motive and purpose would be an outrage quite as brutal.[15]

Such protests were ignored and work on the track continued. Conflicts over financing delayed its completion, but it was serviceable by the mid-1890s. After the turn of the century, the park commissioners decided to replace the track with a multipurpose stadium.[16]

What motivated some of the bonanza kings to invest extensively in raising and racing horses? Although a personal diversion from commercial interests, the sport was also a popular hobby for wealthy businessmen of the Gilded Age. According to a historian of recreation, "ownership of a stable of thoroughbreds became highly fashionable, and the very rich extended their patronage as never before to the turf."[17] Horse breeding and racing enthusiasts in the West included Leland Stanford, who trained trotting horses on the Palo Alto ranch that became the site of Stanford University, and Elias "Lucky" Baldwin, a Comstock mining mogul, whose Santa Anita Rancho in California's San Gabriel Valley contained the largest racing stable in the United States.[18] The sport of kings offered an obvious attraction for the rich, especially since for centuries English monarchs and members of the nobility had owned race horses. Such wealthy easterners as the Vanderbilts, Whitneys, and Belmonts continued that tradition in late–nineteenth-century America. As August Belmont, the millionaire investment banker and father of thoroughbred racing in the United States, tersely remarked, "racing is for the rich."[19] In short, it was a genteel indulgence that allowed the mining nouveaux riches to behave in a manner expected of the wealthy. It also gave the Westerners an occasional opportunity to fraternize with eastern elites, as when Hearst and Belmont watched Ballarat run at Belmont Park in New York. Under such circumstances status-minded Phoebe Hearst probably was less likely to object to the amount of time her husband spent amusing himself "with his *dear* horses."[20]

Yachting was another expensive pleasure of eastern elites in the Gilded Age. Indeed, the period between the Civil War and World War I was something of a golden age for the private ownership of long, beautiful pleasure-craft that decorated the harbors of such fashionable resorts as Newport, Rhode Island, eastern society's summer capital and the site of the annual New York Yacht Club race. Yacht races were a typical feature of the social season among the wealthy just before the turn of the century. By 1888, there reputedly were 120 yacht clubs in the country. By 1913, there were 500 American-owned steam yachts under commission, all measuring over 100 feet. Eighty-one were longer than 200 feet, and 29 exceeded 300 feet. They included the *Alva*, the 285-foot, $500,000 craft of William K. Vanderbilt; the *Lysistrata*, owned by *New York Herald* newspaper heir James Gordon Bennett, Jr.; and the regal *Corsair*, J. P. Morgan's pride and joy.[21]

Although a compelling upper-class status symbol, yachting generally was not popular with the mining barons. In fact, Winfield S. Stratton, the reclusive Cripple Creek millionaire, was called "a fool" by a Denver newspaper reporter "because he didn't come out and buy a fine yacht. . . and have a good time and see life."[22] Few responded to J. P. Morgan's dictum "You can do business with anyone, but you can go sailing only with a gentleman."[23] Since the yachting tradition and the better-known, more established yacht clubs were almost exclusively eastern, western mining entrepreneurs may not have been sufficiently exposed to the avocation to embrace it as their own. While the eastern leisure class owned large steam or sailing yachts capable of going to sea, a western journalist complained in 1891: "Here in California we have no leisure class, to speak of, and our very wealthy men generally 'run to horse.'... Our yachts are smaller, fewer in number comparatively, and are only used on Saturday afternoons and Sundays as a general thing. Moreover, most of our yachting is confined to the bays, the Pacific Ocean in these latitudes being pacific only in name and not in reality."[24]

The few yachting enthusiasts from the mining West pursued the sport in eastern waters. Henry R. Wolcott, who owned the schooner *Fortuna*, was a member of the New York, Larchmont, and American Yacht Clubs of New York as well as the Eastern Yacht Club of Boston.[25] In a 1904 interview after one of his cruises, Wolcott, who was frustrated in his attempts to become Colorado's governor, confessed: "There is nothing in Colorado politics which is as attractive as yachting."[26] Joseph R. DeLamar, who spent his early life as a sailor and in the marine salvage business before turning to mining, was a member of the New York, Larchmont, and Columbia Yacht Clubs during his later residence on the East Coast.[27] Thus, although a few mining moguls who moved east in later life joined the yachting fraternity, most were inclined to seek their pleasure elsewhere.

Their other diversions included the purchase of fashionable transportation vehicles, such as the private railroad cars (usually made by George Pullman) owned by mining barons Darius O. Mills, Samuel Newhouse, and David H. Moffat.[28] However, around the turn of the century, a more novel invention caught the fancy of America's captains of industry—the automobile. Until Henry Ford's mass-production, assembly-line technology made the automobile accessible to ordinary Americans in 1908, it was regarded, often disapprovingly, as a plaything of the rich. However, in 1899, Mrs. Oliver H. P. Belmont tried to give this new toy the dignity of a glamorous competitive sport. On her lawn at Newport she held a concours d'elegance in which such society luminaries as Mrs. Stuyvesant Fish drove flower-laden cars over an obstacle course. According to one student of the rich, "the heyday of the automobile, socially speaking, came in 1904 when William K. Vanderbilt sponsored the Vanderbilt Cup Races, held on a thirty-mile private highway that he and his fellow car buffs acquired on Long Island."[29] However, auto racing did not achieve a truly national popularity until just before World War I with the establishment of the Indianapolis Speedway in 1911."[30]

Despite Vanderbilt's enthusiasm for the new sport, car races were probably too noisy and dusty for most refined society people. As a group the mining elites and their wives generally were too old and set in their ways to readily forsake their stylish carriages for the uncertainty of motorized transportation over roads better suited for horse travel.[31] However, the elegant Samuel Newhouse owned four automobiles and a race car, and some children from mining families caught the automotive fever.[32] Among the entrants at Mrs. Belmont's Newport car rally in 1899 was Mrs. Hermann Oelrichs, the daughter of Comstock millionaire James Fair.[33] Also, his son, Charles, was among the first San Franciscans to buy the new horseless carriage. In 1902, he purchased a Mercedes Special and met his death attempting to set a speed record across the French countryside outside of Paris during a friendly competition with his fellow amateur race driver, William Vanderbilt, Jr.[34] Mining engineer John Hays Hammond recalls that his son and Thomas Walsh's boy were attracted to automobile racing:

> My son Jack and Vinson Walsh acquired automobiles of the same make, in which they used to race each other. In order to keep them within bounds, their parents formed an automobile club, and invited them to join. They felt highly honored at this attention from the elders and ac-

> cepted with alacrity. As soon as the fathers had paid the boys' dues, these same fathers hastily passed a by-law imposing fines and removal of license on any member who was reported for speeding.[35]

Such restrictions were well-advised given the high-spirited behavior of some spoiled children of the rich. However, they failed to prevent Vinson's untimely death in an accident at Newport in 1905 while driving his sister Evalyn's car. She was seriously injured. Later a passing motorist struck and killed her young son near his home.[36] Although a symbol of wealth and privilege, the early automobile brought more pain than pleasure to at least two notable mining families.

Like everyone else, the rich were judged by the company they kept. Men seeking to proclaim their position in society associated with others of comparable status in exclusive fraternal organizations or gentlemen's clubs, thereby continuing a tradition patterned after the English and first begun in this country as early as the eighteenth century.[37] Although most mining leaders belonged to several clubs, there appeared to be one or two favorites in each major western mining region or city that attracted those who lived there. In the privacy of the club rooms, members could enjoy easy conversation, relaxing reading, a friendly game of billiards or cards, and the obvious opportunity to make business deals.

The Montana Club found Charles Broadwater, Samuel Hauser, and Frederick Heinze on its membership list. Incorporated in 1885 for "literary, mutual improvement, and social purposes," the organization occupied impressive quarters in Helena, complete with cherry and walnut furniture, glass chandeliers, Brussels carpets, and marble-topped tables. A bar and billiard, card, and reading rooms with an extensive newspaper file were available for the recreation of the members and their guests, however, no gambling was allowed.[38] In Butte, the pretentious Silver Bow Club, housed in luxurious accommodations on Main Street, was popular with such mine owners as Heinze, Marcus Daly, William A. Clark, and Charles A. Broadwater. It was here that Butte's industrial elite entertained national magnates such as banker J. P. Morgan, James J. Hill of the Great Northern Railway, and Henry H. Rogers of the Standard Oil Company.[39]

In Colorado, the Denver Club was a favorite of Henry R. Wolcott, Thomas Walsh, David H. Moffat, Dennis Sheedy, and Horace Tabor. Wolcott served as president for fifteen years.[40] Moffat, who along with Wolcott and others organized the club in 1880, was a regular visitor. An old friend reported that he "never cared much for society, although he would go when his wife wished him to take her somewhere. It was his custom to spend an hour or so at the Denver Club every evening, and then he would go home."[41] Tabor apparently was an active member. In 1883, he was asked to contribute money to remodel and furnish a new house for the club's use.[42] Later, the club was described as "one of the finest institutions of its class west of New York, the owner of a magnificent building, richly and tastefully furnished and appointed."[43] Although mine owners constituted a large percentage of the membership, ranchers and other members of the business community also belonged, as was true of the Montana Club. The Denver Club was noted for its extensive library, which contained the writings of Hubert Howe Bancroft, Francis Parkman, Thomas Jefferson, and Woodrow Wilson as well as numerous historical biographies, English literature, and some travel books.[44]

In the resort community of Colorado Springs, two other private men's clubs attracted the interest of mine owners. The El Paso Club, established in 1877, claimed to be the oldest men's social club west of Chicago. It was equipped with a dining room, library, billiard room, and card rooms. Irving Howbert, who sat on the board of directors, Thomas Walsh, and Winfield S. Stratton were members. However, the latter seldom visited the club rooms, often preferring to spend his evenings at home in reading or meditation.[45] The leading chronicler of Colorado Springs' history has provided a brief description of the club's activities: "All through the Nineties, full dress was usual at the Club in the evenings even if nothing special was going on. But otherwise the stiff English decorum of the 1880s was modified as more and more Cripple Creek millionaires, who had not won their fortunes by being decorous, joined the Club. Business deals involving huge sums were made over the second floor poker tables."[46] In 1891, outdoor-sports–minded members of the El Paso Club helped form the Cheyenne Mountain Club, which boasted one of the first golf courses in the nation and included mining leaders such as Howbert among its members.[47]

In California, the Pacific Union Club was one of the most exclusive and luxurious gentlemen's social clubs in the state. Incorporated in San Francisco in 1881, it consolidated two older local clubs, the Pacific Club (founded in 1852) and the Union Club (organized in 1854). Albert K. P. Harmon was a member of the latter, and it is likely that he therefore joined the Pacific Union Club, which was housed after 1911 in James Flood's former brownstone mansion on Nob Hill. Other California members included Alvinza Hayward, John P. Jones, John Mackay, and Darius O. Mills.[48] In addition, in 1889, Simeon G. Reed of Portland was elected to membership in the organization. His admission fee and first month's dues amounted to $210.[49] Some bonanza kings in this study who resided or invested in California also belonged to the Society of California Pioneers. More a historical, fraternal organization than a gentlemen's recreational club, it was nonetheless exclusive enough to restrict its membership to males who were in California before 1850 or descended from persons who were there at the time. Founded in San Francisco in 1850, the society assumed the responsibility, which it continues today, of collecting and preserving information on the early settlement of California.[50] Its members included John Q. Packard, who served as vice-president, and William O'Brien, Almarin Paul, and Harmon.[51]

In Virginia City, Nevada, a teeming metropolis by western urban standards of the 1870s, Comstock mine owners frequented the Washoe Club. Housed on the floor above the Crystal, "the millionaires' saloon," the club was described as "surpassing in luxury and in the perfection of its appointments any gentlemen's club in Western America."[52] Its quarters contained a lounge, bar, and cardroom, where John Mackay occasionally held forth during an evening of high-stakes poker.[53]

In Utah, the prestigious Alta Club had a strong following among mining entrepreneurs who lived in Salt Lake City. Founded in 1883 by Enos Wall, with assistance from the Walker brothers, the Alta Club is one of the oldest, most exclusive gentlemen's clubs in the state. Wall served as its first president. Packard was a charter member and such other local magnates as Thomas Kearns, David Keith, Henry Newell, and Samuel Newhouse belonged. The Alta's social prominence was rivaled only by the Commercial Club of Salt Lake City, which was nearly as popular with the same group of mine developers. Newhouse even donated property for the site of the Commercial Club building.[54]

Some western mining entrepreneurs, including those who moved east in later life, found acceptance in the nation's older or more prestigious private clubs. Darius O. Mills, James B. Haggin, and Thomas Walsh belonged to the Metropolitan Club in New York City. Known as the "Millionaires' Club on Millionaire's Row," the Metropolitan was organized by J. P. Morgan in retaliation against Union Club members who had blackballed one of his friends. Mills, Haggin, and Henry Wolcott also claimed to be members of the elite Union Club, founded in New York City in 1836.[55] Although a few mining leaders belonged to such other New York City gentlemen's clubs as the Manhattan and the redoubtable Knickerbocker, the Union League Club, where John D. Rockefeller was a member in good standing, most frequently was cited by those bonanza kings with eastern club affiliations.[56] According to one of William A. Clark's biographers, "compulsive desire for recognition and prestige prompted him to seek a Union Club membership."[57] Other western mining men who joined the club, perhaps for the same reason, included David Moffat, Mills, Wolcott, and Jerome Wheeler.[58]

A few mining moguls belonged to the Rocky Mountain Club of New York City. Known as the "Eastern Home of Western Men," the club attracted western businessmen and politicians as well as such mining entrepreneurs as John Hays Hammond of Idaho, Daniel Guggenheim of Colorado, and Daniel Jackling and David Keith of Utah. Hammond, who served as president, remarked: "It is certainly refreshing to men of Western habits to find the Western scenes and customs reflected in the membership of our Club."[59] Although social ambitions may have prompted some western mining leaders to seek membership in exclusive eastern clubs, others were apparently interested in maintaining their western identity and associations even while visiting or residing in a melting pot like New York City.

Although western and eastern clubs and fraternities provided a pleasurable escape from the scrutiny and responsibility of family life, they attempted somewhat ironically to duplicate in an all-masculine setting many of the comforts and conveniences of home. Regardless of how much time actually was spent at the clubs, they were a badge of social distinction for the wealthy of the Gilded Age and also were useful for the promotion of business interests. Nonetheless, not all bonanza kings were attracted to such organizations. George Hearst said: "I made up my mind that a good man was a good man anywhere, and that there was no occasion for joining any affiliation societies. I made up my mind to be independent, and decided not even to belong to a church."[60] Hearst was an exception. Even in an age that celebrated rugged individualism, most mining entrepreneurs recognized the value of supportive group associations.

Indeed, a number of the mining elites were anxious to secure membership in the most prestigious club of all, the "Millionaires' Club," or U.S. Senate. Of the fifty men, ten actually held seats in the upper house. Others were urged to run but declined, such as John Hays Hammond and John Mackay, who replied: "What, I go to the Senate! I've got no time for that. Besides [Senator] Stewart [of Nevada] is the better man."[61] Still others, such as Thomas Walsh and Adolph Sutro, ran unsuccessfully for election. With the exception of John P. Jones, who represented Nevada for thirty years, the successful candidates served only one term or less and compiled unimpressive records. While little could be expected of freshman senators, their limited prior political experience and brief, lackluster performances suggest that most regarded a seat in the senate more as an honor and costly diversion for the rich than as an opportunity dutifully to serve the public interest or even to become actively involved in political decision making.[62]

Thomas Kearns, William C. Sharon, and James Fair exemplified this pattern. Kearns, who completed an unexpired term from 1901 until 1905, was better known in the senate for being well-dressed than for any legislation he sponsored.[63] According to his biographer,

> the only real debate into which Kearns even entered... was on the subject of mining. . . . His career in the senate was something less than lustrous. He was severely criticized at home for his questionable speech and manners, and some even felt he was "a disgrace" to Utah. To his credit, however, goes his work for Utah conservation, for the opening of the Uintah [Indian] reservation [to

> settlement], and for the raising of Fort Douglas to a regimental post. Beyond the securing of certain appointments, this is about all with which Kearns may be credited. . . . His short time there . . . was, on the whole, uneventful.[64]

Sharon was a particularly disappointing senatorial choice from Nevada. In 1874, he ran a successful but expensive and corrupt campaign against Comstock rival Sutro. He then earned the dubious distinction of having set a new absentee record for the senate by failing to attend sessions in the first year and last four years of his term in office (1875–81). Although his absence in 1875 was due to the financial problems of the Bank of California, he obviously was more interested in the title than the responsibility of high office.[65] In 1880, he ran unsuccessfully for re-election, chiefly against fellow Comstocker James Fair. It has been suggested that the latter spent $350,000 to win the race.[66] According to a contemporary local journalist, "in the entire history of Nevada politics there had never been such a saternalia of corruption as in this campaign. Votes were purchased openly on every side and the price, which in ordinary campaigns had generally ranged at about five or ten dollars, struck the $80 mark for single votes."[67] One scholar claims that "Sharon had the resources to equal 'Slippery Jim' Fair's outlay, but he allowed himself to be outbid, perhaps because he had already held the honor and thus valued it less."[68] Although Fair worked in the senate with his senior Nevada colleague, Jones, to promote the free-silver issue, he quickly tired of politics and parliamentary debate. His biographer asserts that "to be a member of the nation's most eminent millionaires' club was flattering to his ego, but he was soon confessing that the deliberations of that body bored him to distraction."[69] At the conclusion of Fair's term a San Francisco newspaper reported: "About all that can be said for Nevada's junior Senator is that he has had the good sense to sit silent while matters of which he knows nothing are under debate."[70]

George Hearst seemed to agree with Fair's apparent philosophy that first-term senators should be seen but not heard. He seldom spoke in debate and claimed with some pride to be "the silent man of the Senate."[71] In his memoirs, he mentioned his membership on committees concerned with railroads, mining, and Indian affairs and then made the startling admission that he didn't "want to act on these because there is a heap of work in them."[72] Originally, he was appointed to the senate by California Governor George Stoneman to fill the unexpired term of John Miller, who died in March 1886. He served from 23 March until 4 August of that year, but a special session of the Republican-controlled California legislature replaced the Democratic Hearst with a Republican, Abram P. Williams, who completed the remaining seven months of Miller's term.[73] In January 1887, Hearst was elected by the legislature to a full term in the senate. He served until his death in 1891, but his election was tainted by the charge that he bought his seat for half a million dollars and that Chris Buckley, the powerful blind political boss of San Francisco, delivered the votes necessary to his victory.[74]

Like Hearst, Horace Tabor of Colorado filled an unexpired term. After failing to win a full term in the Senate in a hotly contested campaign with his rival, Thomas Bowen, he was offered the remaining thirty days of former senator Henry Teller's term by the Republican caucus of the Colorado State legislature. He took seriously his brief term from 27 January until 3 March, 1883, establishing a good attendance record and introducing two bills in the last week of the session.[75] However, like his mining colleagues, Tabor was much more intrigued with the status and honor than with the responsibility of office. In a letter of 20 February 1883 to his prospective bride, Baby Doe McCourt, he wrote: "Our sessions lasted from 11 am to 2 this morning. That was 15 hours and I do not like very much the office of senator but the honors are great[.] They are next to the President."[76] His interest in the social advantages and prestige of his new position was clearly expressed in another letter to Baby Doe: "I have quite a grand dinner party tomorrow night in fact the best that has been held here [Washington, D.C.] for a long time and I do not know that it has ever been excelled. I have to it the President, Senators Sherman, Morrill and a lot of the best. There will be 20 all together. . . . You watch the associated press despatches [*sic*] sunday morning and you will see an account of the dinner. The room will be hung luxuriantly with flowers and all will be lovely."[77] Shortly thereafter, Tabor, who could never be accused of modesty, wrote Baby Doe that the dinner was the "crowning event of my life politically."[78] A commitment to disinterested public service was not mentioned.

Tabor's Colorado mining colleague, Jerome Chaffee, also served in the Senate for less than a full term. Instrumental in securing statehood for Colorado as territorial delegate to Congress (1871–75), he was rewarded with a seat in the Senate by the Republican members of the Colorado legislature. He served in that capacity for the short term from 1877 until 1879. Although Chaffee was responsible for introducing several bills on mining regulations, territorial representation, and Indian affairs, he initially declined to seek re-election, ostensibly because of failing health.[79] However, he confessed to Nathaniel P. Hill: "I do not intend to interfere or take any part in politics, hereafter, sick or well. It was not my health altogether which induced me to retire from politics. I never took any pleasure or satisfaction in the business and have long wished to get out, and now is a good time and I have a good excuse."[80]

One can only guess at the reasons for his disillusionment. It is possible that having received the ultimate honor of a senate seat, there was no higher public office to which he could aspire. Curiously, he belatedly entered the 1878 senatorial race only to lose to Hill who, according to the *Rocky Mountain News* for 14 September 1880, "put up ten thousand dollars to save the [Republican] party and took a mortgage on the senatorship."[81] Having served until 1885, Hill summed up his experience by observing, "I have never worked harder in my life."[82]

None of the mining elites who ran for or was elected to the Senate was as anxious for the honor as William A. Clark. He clearly had a burning desire to hold the office at all costs. Initially he was chosen U.S. senator in 1890 by the Montana Democrats in the first state legislature. However, the Republican-controlled U.S. Senate seated his Republican rival instead in a contested election that the Montana lawmakers could not resolve. In 1899, after spending $140,000, he was again elected; but Marcus Daly, his political nemesis, provided financial assistance for a Senate investigation of his campaign. When the Senate Committee on Privileges and Elections issued a report charging Clark with bribery, he resigned before taking his seat. Long frustrated in his quest, he finally served in the Senate from 1901 until 1907. His election was facilitated by Daly's declining health, the Standard Oil Company's purchase of controlling interest in Daly's Anaconda Mine, and Clark's political alliance with shrewd and popular Augustus Heinze.[83]

Clark's term in the Senate was undistinguished. He rarely addressed the lawmakers except on matters which concerned him personally. In fact, according to one political biographer, "his activities . . . were not of a national character, but were mostly sectional, even personal."[84] For example, although Clark voted for such national reform legislation as the Pure Food and Drug Act, he opposed, as did other westerners, Theodore Roosevelt's conservation and national forest land withdrawal programs, undoubtedly for fear in part that they would jeopardize his timber and mineral interests in Montana.[85] Like his lavish lifestyle, Clark's political career, on which he reportedly spent $2 million, was largely motivated by a desire for social recognition.[86] His most recent biographer claims that "Clark seemed to relish more the title than the duties of a U.S. senator. . . . He was frequently absent, tending to his business and social concerns, and one imagines that he did not enjoy the necessities of catering to his constituents. . . . For the rest of his long life, though, he insisted upon the title 'Senator.'"[87] Upon his death in 1925, a New York newspaper commented that he had resolved "to add the title of Senator to his collection of objects d'art."[88]

The notion that a Senate seat was an especially important status symbol for socially ambitious mining kings is suggested by the political activities of Thomas Walsh. He once told his friend John Hays Hammond that he would "like to play around in politics," and his decision to move to Washington, D.C., rather than to another eastern city, most likely was in part politically motivated.[89] In 1899, some Colorado friends urged him to seek a congressional seat, but he refused.[90] By 1906, he had become more receptive to the idea of public office and seemed to regard a position in the Senate as the titular prerogative of a successful business leader. He explained that a U.S. Senate seat would give him an "official character."[91] However, the Colorado legislature chose Simon Guggenheim, the state's "smelter king," for the honor.[92]

Although Walsh's campaign failed, his aspiration to join the exclusive fraternity of other big businessmen who sat in the so-called "Millionaires' Club" was not unusual. Many of the mining leaders either sought or were urged to seek the honor and prestige of a seat in the Senate. This generally meant the expenditure of huge sums of money and questionable or improper campaign tactics. Before ratification in 1913 of the Seventeenth Amendment to the U.S. Constitution, state legislatures chose the senators. Mining magnates, like other wealthy individuals, often bought influence with the voters and lawmakers to get themselves elected. The amoral politics of the period is perhaps best reflected by Senator John P. Jones's alleged observation that "in a

Nevada Senatorial campaign a million might be spent and no man corrupted."[93]

Although high public office was important to some, the grand tour of the Old World was a more socially significant and obligatory experience for the nouveaux riches of the Gilded Age. Therefore, virtually all the mining elites and their families spent some time abroad, ostensibly in search of the cultural sophistication that was lacking in raw industrial America. Sandy Bowers reflected the social compulsion behind foreign travel when he boldly and ironically announced: "I've got money to throw at the birds. Thar ain't no chance for a gentleman to spend his coin in this country, and thar ain't nothin' much to see, so me and Mrs. Bowers is agoin' to Yoorop to take in the sights. . . to see the Queen of England and the other great men of them countries."[94] The naive, presumptuous Bowers was unable to meet the queen or other royalty for that matter, but he nonetheless satisfied an important expectation of the American upper class. As a contemporary observer reported, "somebody had told him that the tour of Europe was the correct thing for wealthy Americans, and he made up his mind to do it or die."[95]

Foreign travel itself was not just an important elitist custom. It gave western nabobs the practical opportunity to fulfill such other socially sanctioned forms of elite behavior as book and art collecting and to buy expensive furnishings for their American mansions. Elizabeth "Lizzie" Mrs. Ralston (née Fry), returned from Europe in 1869 with $25,000 worth of furnishings for her Belmont estate.[96] Darius O. Mills and his wife partially furnished their Milbrae mansion with decorations they acquired at the international Paris Exposition of 1867.[97] Later, in the 1880s, Joseph Walker and his wife, Mary Ann, took several trips to Europe, on each occasion returning with a large collection of paintings and sculpture to embellish their exquisitely furnished mansion in Salt Lake City.[98]

At the same time the grand tour of the continent was not entirely a socially conditioned response to certain role expectations of the rich. Some bonanza kings had sincere personal reasons for travelling abroad. Allegedly, James B. Haggin's extensive foreign tours, including a trip to the Holy Land, were devoted more to serious study and information gathering than to recreational sight-seeing or frivolous art collecting.[99] Some mining entrepreneurs used the grand tour to visit their birthplace or ancestral home. In 1899, the Walshes made the first of several journeys to the Old World, touring England, France, Switzerland, Austria, Bavaria, and, of course, Tipperary, Ireland, the county of Walsh's birth. Later his daughter claimed that for her father "the whole of our journey was simply part of a pilgrimage to Tipperary."[100] Simeon Wenban, a prominent developer of silver mines in eastern Nevada, visited his childhood home, Hawkhurst, County Kent, England, during his European sojourn.[101] At least one mining leader, Andrew J. Davis of Montana, sought to improve his health by going abroad in 1882.[102] Others and their wives crossed the Atlantic to see family members. John D. Fry, William C. Ralston's father-in-law and a member of the Comstock's so-called "Bank Crowd," went overseas in 1890 to visit his son.[103] Joseph R. DeLamar travelled to Holland to see members of his family before returning from his European tour to "God's Country" in the fall of 1891.[104] Amanda Knight's trip abroad included an important stop in England to spend time with her daughter, Inez, who was doing missionary work for the Church of Latter-Day Saints.[105]

Religious, political, and business interests also help explain why the mining moguls went abroad. A devout Catholic, Thomas Kearns highlighted his European excursion in 1901 with an audience with Pope Leo XII, whose blessing he received.[106] John P. Jones travelled to Europe on two occasions to study the international monetary situation as U.S. senator and chairman of the U.S. Monetary Commission.[107] Although Winfield S. Stratton was ill three of the six months he spent abroad, he remained on the continent during the negotiations leading to the sale of his Independence Mine to British interests in 1899.[108]

Since only a few recorded their impressions, it is difficult to know how the mining entrepreneurs as a group regarded Europe. Those who made observations were more negative than positive, suggesting that the journey abroad for most was apparently motivated more by class custom than by a personal interest in European culture. In 1878, Simeon G. and Amanda Reed visited Ireland, England, Scotland, Germany, Switzerland, Italy, and France. Amanda noted that they saw the major tourist attractions, stayed in the most exclusive hotels, and attended plays and concerts.[109] However, according to Simeon's biographer, "the knowledge and polish which could be derived from the European tradition of wealth and leisure failed to interest him. Trout fishing, duck hunting, or watching a good horse run gave him more pleasure."[110] He complained: "About the first thing I shall buy after getting Home will be an American flag. It will be such a comfort to look at it occasionally and thank your lucky star that you were born under its ample folds."[111] After his trans-Atlantic trip, Kearns reported: "I tell you, there is nothing which makes an American appreciate his country like a trip abroad. These European people are simply not in it at all with us, either in a business way or in the enjoyment of liberty and opportunity. We have them out-classed on every hand, and I return more convinced that we live in the only country on earth."[112]

Horace Tabor visited London, Paris, and Marseilles in 1889, but returned home unimpressed. In a dictated interview with a member of Hubert H. Bancroft's staff, he argued: "Here we have got our twelve and twenty feet of soil all over these great prairies. Pretty near all you can see to admire over there [Europe] is old tumbled down buildings."[113] On the other hand, the better-educated Nathaniel P. Hill, a professor of chemistry before turning to mining, was more discriminating and favorable in his appraisal of the Old World than was the unlettered Tabor. Although he found London dirty, he wrote his wife in 1866: "The only objects in London which are really worthy of much notice are those which derive their interest from their age and historical associations. If my object here was sightseeing and you were here to enjoy it with me I could spend a month very happily."[114] Later he distinguished between the cities he had visited: "A more perfect contrast than that which exists between Paris and London cannot be imagined. Paris is gay, bright, cheerful. The streets are beautifully paved and kept sumptuously clean. The whole world of Paris is in the street."[115]

Despite the mixed reactions from those who left observations of Europe, one who never made the grand tour apparently regretted it. A year before his death, George Hearst wrote: "I never travelled in Europe myself, but sent my family over there, at one time. I won't say that I would not like to have gone there, and I don't know but that I ought to have gone, as I believe I might have learned a good deal over there."[116] Nonetheless, in 1885, Phoebe Hearst complained: "Mr. Hearst will never cross the ocean with us. Even if he should, he would disapprove of all we might enjoy."[117] Perhaps advancing age mellowed Hearst's outlook on the Old World.

Regardless of their husbands' opinions of Europe, the grand tour of the continent served a very important purpose for the wives of some American nabobs. It provided the opportunity to make contacts in European and especially British society in hopes of achieving the ultimate social triumph of finding foreign titles for their marriageable daughters. As European aristocratic fortunes declined around the turn of the century, members of the Old World nobility were more eager than ever to make suitable international matches with wealthy American heiresses. Resulting marriages included those of May Goelet to the Duke of Roxburghe, Ethel Field to Earl Beatty, Mary Leitner to Lord Curzon, and Anna Gould first to Count Boni de Castellane and then to his cousin, the Duc de Talleyand-Perigord. Perhaps best-known was the union of Consuelo Vanderbilt, William K. Vanderbilt's daughter, and the Duke of Marlborough. Like an indomitable stage mother, Consuelo's mother, Alva, orchestrated the affair against her daughter's own wishes, but to the financial benefit of the Marlboroughs.[118]

Nonetheless, few daughters of the western mining leaders followed the example of their better-known, title-seeking, eastern counterparts. Most married members of the American upper class. Occasionally, this meant merging western money with eastern wealth or prestige. Such was the case when Darius O. Mills's daughter, Elizabeth, married Whitelaw Reid in 1881. Publisher of the powerful *New York Tribune*, Reid was the Republican vice-presidential candidate in 1892 and later was U.S. ambassador to Great Britain.[119] The marriage of Fannie Josephine, Jerome Chaffee's daughter, to U. S. Grant, Jr., was another celebrated east-west connection of the period. In later life, Chaffee spent much time at his daughter's home, Murryweather Farm, Westchester County, New York, and died there in 1886.[120] The most publicized union of eastern wealth and western mining money involved James Fair's two daughters. In 1890, Theresa, the eldest, married Hermann Oelrichs, a member of a prominent New York and Newport family. Her father, who earlier had been accused of "habitual adultery" by his former wife, was not invited to the ceremony. In 1899, his other daughter, Virginia, wed William K. Vanderbilt, Jr., the commodore's grandson and Consuelo's brother.[121]

Despite the prevailing tendency for the children of western mining leaders to marry fellow Americans of the same general class background, some mining daughters found aristocratic European suitors. In 1880, Florence "Flora" Sharon, the daughter of William C. Sharon of Comstock fame, became the wife of Sir Thomas Hesketh, an English baronet. Thereafter, Lady Hesketh lived abroad, although she made the long trip from Europe to visit her ailing father in San Francisco shortly before his death in 1885.[122] Helen Kearns also found a European husband of high rank. Her second marriage was to Mario Enzo Gabellini, a member of the Italian nobility.[123] Eva Mackay, John Mackay's adopted daughter, also was attracted to an Italian. In 1885, she married Ferdinand Julian Colonna, Prince of Galarto, a member of the Neopolitan branch of this distinguished Italian family. Later, when the marriage failed, Mackay arranged a considerable financial settlement to secure his daughter's divorce. According to his biographer, "it is probable that this expenditure gave him a certain satisfaction, for the custom of marrying off rich American girls to impoverished holders of foreign titles seemed to him the height of folly."[124] During her foreign travels, Evalyn Walsh was linked romantically with a variety of titled gentlemen, including Italian noblemen. In 1905, the Italian Prince Altieri, apparently seeking a rich American heiress, pursued her. Although amused and flattered, she decided against the marriage.[125] Her decision might have been influenced by the objection of her brother, Vinson, who, in a letter to his mother, protested: "Don't let Sis have anything to do with the prince. I am afraid from your letters that you have been blinded by the thought of a title and are thinking about the high position it would give you rather than of Sister's happiness and if you have not yet influenced her one way or the other start right in and get the idea of marrying that dago right out of her head."[126] In addition to exposing a nativistic prejudice against Italians during an era of heavy emigration from Italy to the United States, the letter clearly reveals a mother's desire to secure social respectability in the world of the American nouveau riche by marrying her daughter to a foreign nobleman. Although the mining wives were apparently less active in this effort than were other wealthy American women, there were many presumably impoverished European aristocrats willing to oblige them and their daughters, especially if the price was right. Indeed, according to one estimate, by 1909, "five hundred American heiresses had been exported for the improvement of the family name, along with $220 million."[127] Evalyn Walsh was not among them. In 1908, she married Edward B. McLean, whose multimillionaire father owned the *Washington Post* and the *Cincinnati Enquirer*, and began a troubled relationship that may have made her wish for the earlier blandishments of European aristocracy.[128]

As were American millionaires generally, the mining moguls were not inclined to withhold the material benefits of wealth from their wives, most of whom apparently were given sufficient money to indulge their various whims and interests, whether for clothes, jewelry, foreign residence, or society parties. George Hearst epitomized this indulgence when he wrote to Phoebe: "Spend as much money as is necessary for your pleasure and not think of it. For if you are to feel that you are doing wrong you lose all your pleasure. . . . So do not write me that you are feeling bad about what it costs. You will get all you want for your pleasure and comfort and it only makes me hapy [*sic*] for you to have the money to use in that way."[129] Thomas Walsh apparently shared Hearst's attitude. His daughter recalls that "if Mother so much as breathed a wish for anything—even a set of $1,800 furs from Gunther's—she got it, overnight at the latest."[130] Horace Tabor married Baby Doe McCourt in a Washington, D.C., society wedding with President Chester Arthur as the guest of honor that has been described as "the most lavish money could buy."[131] Among the gifts he presented his new bride was a $75,000 diamond necklace.[132] According to his biographer, Samuel Newhouse "spent a great deal of money on his wife. She had some of the costliest jewels of any lady of her time."[133] William C. Ralston reportedly gave his wife jewels and gowns in "faultless taste and plenty," including the latest Paris fashions.[134] A society columnist for a Denver newspaper filed this report of Alice Hill's extravagant attire at a White House reception in 1881: "Mrs. Hill wore one of the most elegant costumes seen this winter at the capital.

It was of brocaded velvet, of a shade between plum color and brown, combined with satin de Lyon of the same shade. A small hat matched the color of the robe. Over the top was carelessly caught some fine white lace. Diamond ornaments completed the costume."[135] Although the mining wives had minds and serious interests of their own, they were often encouraged to appear as ornamental showpieces befitting the success of their rich husbands. If nothing else, such fashion and flare made for interesting reading in the society columns of the day.

A few husbands were willing to finance their wives' efforts to achieve prominence in European society even if this meant extended matrimonial separations. John Mackay lived in hotel suites in San Francisco and New York for thirty years while his wife pursued her expensive social career as a hostess in Paris and London. According to his biographer, "he good-naturedly expended huge sums to provide his wife with the lavish settings her nature craved."[136] This included the elegant ball she gave in honor of Mr. and Mrs. Ulysses S. Grant at the Mackays' enormous Paris mansion in 1877.[137]

Samuel and Ida Newhouse seldom lived together. Bored with life in Salt Lake City, Ida spent most of her time abroad seeking the attention of British society. Her husband supported her social ambitions by placing "her under the charge of a tutor who taught her court etiquette and graces then admired in court circles. He then introduced her to the court and the next step was inevitable. She quickly became a royal favorite and thereafter Samuel's most reliable access to the court."[138] Although her position was undoubtedly useful when he sought financial backing abroad for his various enterprises, it was obviously an ego fulfillment for his wife, who was popular with the Prince of Wales, later the philandering Edward VII, king of England.[139]

It should not be assumed that all mining barons were as generous to their wives as the aforementioned individuals or that all the wives were eager for aristocratic distinctions and materialistic embellishments. For example, although Amanda Knight attended a tea at Buckingham Palace given by Queen Victoria, she was not known for her aristocratic pretensions or conspicuous consumption. Committed to the values of home, family, and church, she was active especially in Mormon charity work as president of the Fifth Ward Relief Society and a member of the Utah State Relief Society Board.[140] Thomas Kearns's wife, Jennie, endowed the Saint Ann's Orphanage in Salt Lake City, which met the needs of seventy motherless children. She visited the facility regularly and provided beautifully decorated trees and presents for the children at Christmas.[141] It has been said of Leah Sutro (née Harris) that "she had no dreams of grandeur, and would have been most happy to see her life revolve around home, husband, children, synagogue, and the charities of the congregation. Whenever she could, these were the things she did, but as her husband's life expanded she assumed the roles he forced upon her with dignity and charm," especially that of hostess.[142] A noted social worker and reformer on behalf of the rights of working women, Natalie Hammond, who followed her famous mining-engineer husband into many remote areas of the world, was inclined to help others more than herself. Reflecting upon his wife's good works, John Hays Hammond recalled:

> A woman . . . who has sympathy and understanding can do much to ensure the co-operation of her husband's staff. My wife learned first aid when she was a young woman, and many times she applied that knowledge and afforded great relief to the suffering when physicians were not available.[143]

> I remember that my wife was continually engaged in work of civic and philanthropic nature. . . . She would find something worth remedying, raise the money needed, and administer it personally. She never stopped

> in the middle of an undertaking. If her interest was a hospital or a charitable home, she would frequently drop in and talk to the doctors, to the directors, and to the inmates so that there was always the direct personal touch.[144]

Natalie Hammond was instrumental in the work of various welfare organizations. In 1911, she founded and became national chairwoman of the woman's welfare department of the National Civic Federation. Under her leadership, the federation initiated factory inspection, night schools, dispensaries, medical clinics, and classes in domestic economy for the benefit of female employees in many industrial areas. She also was a founder and vice-president of the Women's Institute of London, which served the interests of working women, and was president of the women's clinic for working girls in Washington, D.C. During the infantile paralysis epidemic of 1916, she organized the Militia of Mercy, which brought aid to victims of the disease among the poor. During World War I, she inaugurated the Children's Christmas Fund and sent gifts to the children of the warring nations.[145] Her concern for war refugees was shared by other mining wives, such as Carrie Reed Walsh, who converted part of her Washington home into a workshop for reconditioning and manufacturing clothes for the women and children of Belgium and France. Although Walsh employed a staff of seamstresses, she contributed her own labor to the project.[146] While there were other self-sacrificing philanthropists among the mining wives, it is virtually certain that they could all look forward to a comfortable self-indulgent life befitting proper Victorian ladies of their social position. This included for their children educational opportunities that they and their husbands often lacked.

Given their commitment to educational philanthropy in particular, it is not surprising that many mining elites and their wives were concerned about the schooling of their own offspring. Some were exposed to the education of travel and formal instruction abroad not only as a result of the popular tour of the continent, but because the women occasionally took additional trips abroad with their children and resided in Europe for extended periods. In 1867, Lizzie Ralston left the country to attend the Paris international exhibition and allegedly to acquire a French tutor for her three children.[147] In the early 1870s, Phoebe Hearst spent a year and a half visiting the museums, art galleries, and cathedrals of the Old World with her ten-year old son, William Randolph Hearst.[148] According to Cora Older, the wife of Fremont Older, the longtime editor of one of William Hearst's San Francisco newspapers, "into the ten-year old mind of her son she poured the best of herself and the best of all that there was in Europe."[149] According to his biographer, William A. Clark also wanted to give "his children the advantages of a European education," hence his first wife, Katherine, and the children lived several years in Europe, where their father, a self-conscious Francophile, introduced them to French culture.[150] John D. Fry also preferred that his son study abroad. Robert Fry received formal instruction in Paris from a private tutor.[151]

European colleges and universities attracted some children of the bonanza kings. William C. Sharon's son, Frederick, a Harvard graduate, took courses in Paris and at Beaumont College in England.[152] Emma Sutro, one of Adolph's four daughters, pursued her medical studies in Paris and at the Toland Medical School in San Francisco (which merged with the University of California medical facility in 1873) before becoming a physician.[153] Charles Broadwater, Jr., the son of Charles Broadwater, Sr., one of Montana's leading mining capitalists, also studied abroad. He received a B.A. degree in 1899 at the University of Berlin and an M.E. degree in 1904 from the Royal School of Mines, University of London. Later, he became a nationally prominent mining engineer and oil products industrialist.[154]

Most children from mining families, however, attended distinguished private schools and colleges in the United States. The Walshes took care in providing their children with a proper education at exclusive eastern institutions. In 1904, Thomas Walsh, who had little formal education, wrote to his son about the advantages of attending a noted academy: "We have had constant inquiries about the differrent [*sic*] schools and believe that the Laurenceville School is far and away the best for you both for its high standing at Yale[,] the fine class of boys who are going there[,] a great many of whom are sons of our friends and the good moral air that prevades [*sic*] it. . . ."[155] Hopefully, young Walsh was able to improve upon his father's limited command of the English language. The effort to turn his sister, Evalyn, into a "decorous lady" included attendance at dancing school, the supervision of a French governess, and enrollment at the Mount Vernon Seminary and the Holton Arms. The latter she described in her memoirs as "very small, very select."[156]

Nathaniel P. Hill's two daughters attended prominent eastern finishing schools, Miss Graham's schools in New York City and in Farmington, Connecticut. His son graduated from Brown University and then followed his father into the mining business.[157] Despite William A. Clark's apparent obsession with French culture, his children from his two marriages also were educated on the East Coast by private tutors or at exclusive finishing schools, such as Miss Spence's School in New York, where a daughter from his second marriage, Huguette Marcell Clark, received her education. Another daughter, Mary Joaquina, was given the advantage of a Fifth Avenue finishing school, "the well-known institution directed by Miss Reed." While W. A. Clark, Jr., earned a law degree from the University of Virginia in 1899, Charles Walker, his elder brother, received his early training from private tutors and went on to receive a degree from Yale.[158] The latter distinction he shared with John Hays Hammond's son, John Hays, Jr., who graduated from his father's alma mater and later became a prolific inventor of important electronic systems and devices, including those essential to the modern space program.[159]

Thomas C. Power, Montana's U.S. senator in the 1890s, with investments in steamboats, mines, cattle ranches, and banks, had one son, Charles Benton, who graduated from Georgetown College (later University) in Washington, D.C., and received a law degree from Columbia College in New York.[160] Irving Howbert also sent his children east for their college education. His daughter, Alice, attended Vassar, while his son, William, studied at Yale.[161] Among those who chose higher education in the East, there was apparently a definite preference for long-established or Ivy League institutions.

Nonetheless, some bonanza kings decided to educate their children in western schools rather than in the more fashionable eastern colleges. Thomas Kearns sent two of his sons, Edmund and Thomas, Jr., to Santa Clara College in California.[162] Given their Mormon upbringing and Jesse Knight's financial support of the institution, it is not surprising that his children were educated at Brigham Young Academy (now University) in Provo, Utah.[163] Although Rosa Sutro, one of Adolph's daughters, attended Miss Reinhardt's English and German School in Baltimore, she eventually enrolled at Mills Seminary (later Mills College) in Oakland, California.[164]

It should not be assumed that all the mining children made the most of the educational opportunities provided by their parents' wealth and devotion. Some squandered their good fortune, as when one of William Randolph Hearst's many practical jokes earned him an expulsion from Harvard during his junior year in 1885.[165] Similarly, Henry Newell's adopted son, Willie, failed to take advantage of his education. Disinterested in school and in constant trouble, he felt that his father's wealth allowed him the luxury of irresponsibility. By the time he died in 1925, he had been married five times.[166] Although there were rebels, miscreants, and idlers among the children of the mining rich, those who wanted the privilege and prestige of an expensive, usually private, education generally found their parents willing to indulge and encourage them.

Parental indulgence also took the form of material gifts and financial subsidy. A few examples illustrate how far some bonanza kings were willing to go to satisfy and support their children. William A. Clark endowed his sons, Charles and William, Jr., with money enough to allow them to enjoy their expensive habits and hobbies.[167] James Fair tried to help his son Charles overcome alcoholism by rewarding his rehabilitation efforts with notable additions to his stable of race horses. However, the ungrateful young Fair resumed drinking and enraged his father by marrying a San Francisco brothel owner.[168] Some mining fathers indulged their children with houses rather than horses. In honor of his daughter's marriage to Whitelaw Reid, Darius O. Mills presented the happy couple with a mansion on Manhattan's Madison Avenue.[169] In 1899, John Mackay purchased a fifty-room Louis XIII mansion, Harbor Hills, at Roslyn, Long Island, and gave it to his son Clarence as a gift on the occasion of his marriage to Katherine Duer of New York.[170] Thomas Walsh generously indulged his only daughter's extravagant tastes for clothes, carriages, jewels, and a red Mercedes automobile. His wedding present to Evalyn was the $120,000 necessary to buy one of the world's most famous diamonds, the 92.5 carat Star of the East from Cartier's in Paris. However, she is best remembered for possessing the more celebrated and controversial Hope Diamond.[171] Perhaps the Hearsts were responsible for the most memorable acts of parental indulgence among the mining families. Following his son's departure from Harvard, George Hearst gave his financially troubled *San Francisco Examiner* to his only child, thereby creating the opportunity for a controversial career that would influence American journalism and politics. William Randolph Hearst's newspaper empire was substantially aided thereafter by the millions of dollars George left his wife, Phoebe, upon his death in 1891.[172] It has been suggested that philanthropic Phoebe's "greatest single charity was probably her son, to whom she 'loaned' $10 million during her lifetime and left an additional $11 million."[173] Even for the publicly generous Phoebe, charity began substantially at home.

Other examples of parental generosity could be cited, but these epitomize the general pattern of indulgence within most of the mining families. At the same time, some parents were apparently fearful that excessively spoiled children and grandchildren would become unproductive, irresponsible adults. Lloyd Tevis, a leading mining investor and president of Wells, Fargo and Company from 1872 until 1892, was aware of the disadvantages of a pampered life when he wrote his daughter Louise about improving the behavior of her son, John, by employing him on the Haggin-Tevis cattle ranch in New Mexico:

> The life there is comparatively free from dissipation of any kind—is hard but not unhealthy; and the men though rough specimens of humanity, are supposed to be honest and courageous—and I know of no place, and no occupation where a pretentious or foolish young man would be so apt to have the conceit and nonsense "knocked out of him"! Your brother Hugh lived there over seven years on such a ranch, and among such men, participating in all their duties and hardships— On his return from his school East your son Lloyd did the same thing on this same ranch of ours in New Mexico, for six months, and reported it no doubt truthfully as the most valuable experience of his life. His great and most radical change for the better, commenced from that date.[174]

The elder Tevis clearly expressed his concern about the negative effects of wealth on the character of his descendants when he earlier wrote his daughter about her other son, Lloyd Tevis Breckinridge: "You will find him an extravagant young fellow, and I particularly advise you against permitting him to throw away money on useless things! I will arrange to furnish him with a regular allowance as soon as we can ascertain what it should be."[175] Although few, if any, of the bonanza kings would have agreed with Andrew Carnegie's philosophy about the value of virtually disinheriting one's offspring, some undoubtedly recognized that kindness and generosity had moral limits, especially when raising children.

It would appear that most of the bonanza kings took pride and pleasure in the leisure activities and luxuries their great wealth permitted, even if many of their interests were dictated more by social custom than personal choice. However, considering their long illustrious business careers and the fact that most had risen to affluence from lower- or middle-class origins, it is probable that the mining business itself was their greatest pleasure. Certainly, some had difficulty separating their work from their social lives. A letter from Jesse Knight to his son Raymond testifies to the problem: "I know you are considerably like me; put business ahead of pleasure always and simply make slaves of ourselves for business. But I suppose this is a pleasure [and] rather than anything else the reason why we do it."[176] David H. Moffat apparently shared this view. In 1900, when a visitor suggested retirement, Moffat replied: "I am not satisfied to do nothing. I want to be doing something new—building or developing. That is the way I get my pleasure out of life."[177] John Hays Hammond may have had trouble putting business aside when not on the job. Therefore, during Sunday luncheons with his staff, his wife, Natalie, imposed restrictions on conversation. "She established a rule of fining anyone who attempted to discuss technical matters. If 'pump,' 'gear,' 'shaft,' or any other such term was used at our home, she would fine the person using it, the money to be given to charity. Because it was difficult at first to teach the men not to refer to the subject of mining, she collected quite a sum."[178]

For mining investors such as Tevis, who confessed that "I have more money than I know what to do with," their wives were willing to help direct the family fortune into such socially proper channels as European tours, society parties, select schools for the children, worthy charities, and extravagant wardrobes and personal accessories.[179] The mining moguls participated in these diversions and decisions but many probably found greater enjoyment at the horse races or the poker tables of their favorite clubs. Although their luxurious status symbols reflected in part the social ambitions of their wives, the mining men as a group spent money on recreational, political, and family activities that were made fashionable by the lifestyle of better-known, eastern business elites. While they occasionally deviated from the norm, as in the infrequent ownership of yachts, most conformed to the Gilded Age standards of conspicuous consumption clearly defined by the precedent and publicity of eastern elitist behavior.

NOTES

1. Godkin, "The Expenditure of Rich Men," 119.

2. Thorstein Veblen, *The Theory of the Leisure Class* (New York: Modern Library, Inc., 1934 [orig. pub. 1899), 264.

3. Donald J. Mrozek, *Sport and American Mentality, 1880–1910* (Knoxville: University of Tennessee Press, 1983), 103–4.

4. John R. Betts, *America's Sporting Heritage: 1850–1950* (Reading, Mass.: Addison-Wesley Publishing Co., 1974), 142–49. See also Foster R. Dulles, *A History of Recreation: America Learns to Play* (New York: Appleton-Century-Crofts, 1965), 240–41.

5. See Harriet Miller and Elizabeth Harrison, "Marcus Daly's Western Champions," *The Western Horseman* (March 1958): 26–27; Malone, *The Battle for Butte*, 81; and Glasscock, *The War of the Copper Kings*, 162–65.

6. Malone, *The Battle for Butte*, 81. See also Glasscock, *The War of the Copper Kings*, 164–65.

7. As quoted in Shoebotham, *Anaconda, The Life of Marcus Daly*, 153.

8. Johnson et al., eds., *Dictionary of American Biography*, 8:84; *New York Times*, 13 September 1914; *San Francisco Chronicle*, 27 March 1966. The *Alta California*, 2 October 1885, reported the shipment of Haggin's yearlings from Kentucky to California.

9. Older, *George Hearst*, 173–76. See also Swanberg, *Citizen Hearst*, 52–53. Despite her reservations about horse racing, for evidence that Phoebe Hearst actually attended one of the premiere races in the country, the Suburban, in the company of "Mr. Clark" (possibly William Andrews Clark), see letter, P. A. Hearst to "My dear little girl" (Miss Janet Peck), undated, Orrin M. Peck Papers, Henry E. Huntington Library and Art Gallery, San Marino, California.

10. Reed Papers, 23:83, 356.

11. Reed to Martin Winch, 18 April 1892, 34:121, ibid. See also 31:56.

12. Reed to L. C. White, 9 November 1893, 37:27, ibid.

13. Brechin, "A Little Place in the Country," 14; Lewis, *Silver Kings*, 228; and Lavender, *Nothing Seemed Impossible*, 238, 251, 248–49.

14. Waters, *Midas of the Rockies*, 149–50.

15. Quoted in Raymond H. Clary, *The Making of Golden Gate Park: The Early Years, 1865–1906* (San Francisco: San Francisco Examiner, 1980), 72–73.

16. Ibid., 83, 97, 107–8, 128–29, 149–57.

17. Dulles, *A History of Recreation*, 240.

18. Brief biographical sketches of Stanford and Baldwin, noting their horse-racing interests, are available in Hart, *A Companion to California*, 26, 423. Additional information on Stanford's equestrian affairs can be found in Oscar Lewis, *The Big Four: The Story of Huntington, Stanford, Hopkins, and Crocker, and of the Building of the Central Pacific* (New York: Alfred A. Knopf, 1938), 173–75. Baldwin's career can be traced in Carl B. Glasscock, *Lucky Baldwin, The Story of an Unconventional Success* (Indianapolis: Bobbs-Merrill Co., 1933).

19. Quoted in David Black, *The King of Fifth Avenue: The Fortunes of August Belmont* (New York: Dial Press, 1981), 289.

20. See Older, *George Hearst*, 174–75. For the comments of Phoebe Hearst, see letter, P. A. Hearst to "My dear little girl" (Janet Peck), undated, Orrin M. Peck Papers.

21. Jacqueline Thompson, *The Very Rich Book* (New York: William Morrow and Co., 1981), 246–47; Thorndike, *The Very Rich*, 112, 271; and Betts, *America's Sporting Heritage*, 150–51.

22. As quoted from the *Denver Post* (undated) in Waters, *Midas of the Rockies*, 149.

23. As quoted in Mrozek, *Sport and American Mentality*, 123.

24. As quoted in Betts, *America's Sporting Heritage*, 150–51; originally from the *Overland Monthly* 16 (June 1891): 562.

25. Dawson, Scrapbooks, Public Men, I, 61:331; and Hall, *America's Successful Men of Affairs*, 2:888.

26. *Denver Republican*, 7 June 1904.

27. *National Cyclopedia of American Biography*, 28:257–58.

28. Newhouse's "personalized, plush railroad car" is mentioned in Rudd, "Samuel Newhouse," 303. Moffat's private car, *Marcia*, was named for his daughter. See Mehls, "David H. Moffat, Jr.," 240.

29. Thorndike, *The Very Rich*, 18. See also Dulles, *A History of Recreation*, 312–18.

30. Betts, *America's Sporting Heritage*, 83.

31. For information on the carriages and horses preferred by Denver's elite, including Mrs. N. P. Hill and Mrs. David Moffat, see the *Rocky Mountain News*, 20 July 1879.

32. Rudd, "Samuel Newhouse," 303.

33. Thorndike, *The Very Rich*, 18.

34. Lewis, *Silver Kings*, 208–9.

35. Hammond, *Autobiography*, 2:485.

36. McLean, *Father Struck It Rich*, 124–26, 238.

37. Thompson, *The Very Rich Book*, 193.

38. A description of the club is in Lewis Atherton, *The Cattle Kings* (Bloomington: Indiana University Press, 1961), 67–68. See also the *Helena Weekly Herald*, 16 April 1885; the *New York Times*, 5 November 1914; and the *National Cyclopedia of American Biography*, 45:175.

39. McNelis, "The Life of F. Augustus Heinze," 24; Shoebotham, *Anaconda, The Life of Marcus Daly*, 97; Malone, *The Battle for Butte*, 73; the *New York Times*, 5 November 1914; and the *National Cyclopedia of American Biography*, 45:175.

40. Hall, *America's Successful Men of Affairs*, 2:888.

41. *Denver Post*, 19 March 1911.

42. Letter, A. A. Denman, Committee Secretary, to Tabor, 22 August 1883, folio file 23, no. 8, Tabor Papers.

43. Hall, *History of Colorado*, 4:605.

44. Atherton, *The Cattle Kings*, 68.

45. Byers, *Encyclopedia of Biography of Colorado*, 248; *Who's Who in*

America, 1930–31, 16:1143; the *Denver Republican*, 9 April 1910; and Marshall Sprague, *Newport in the Rockies: The Life and Good Times of Colorado Springs* (Chicago: Sage/Swallow Press Books, 1980), 226. See also George H. Webster, "El Paso Club of Colorado Springs," *Colorado Magazine* 26 (April 1949): 104–12.

46. Sprague, *Newport in the Rockies*, 227.

47. Ibid., 227–31; and *Who's Who in America, 1930–31*, 16:1143.

48. Angel, *History of Nevada*, unnumbered pages between 116 and 117; Hart, *A Companion to California*, 318. See also various editions of the San Francisco blue books or elite social directories. The original was *The Elite Directory of 1879* (San Francisco: The Argonaut Publishing Co., 1879).

49. Reed Papers, 23:83, 356.

50. A brief description of the Society of California Pioneers is available in Hart, *A Companion to California*, 413; and Kevin Starr, *Americans and the California Dream, 1850–1915* (New York: Oxford University Press, 1973), 110.

51. *Oakland Enquirer*, 30 April 1896; Johnson et al., eds., *Dictionary of American Biography*, 13:621; Leigh H. Irvine, ed., *A History of the New California, Its Resources and People*, vol. 2 (New York: Lewis Publishing Co., 1905), 2:673; and Guinn, *History of the State of California*, 363.

52. As quoted in Lewis, *Silver Kings*, 26.

53. Ibid., 6, 53.

54. See Orvin N. Malmquist, *The Alta Club, 1883–1974* (Salt Lake City: privately printed, 1974); The Alta Club Papers, 1883–1984, Marriott Library, University of Utah, Salt Lake City; Press Club of Salt Lake, *Men of Affairs in the State of Utah, A Newspaper Reference Work* (Salt Lake City: Press Club of Salt Lake, 1914), unnumbered pages; *National Cyclopedia of American Biography*, 16:411; 20:145; Larsen, "The Life of Thomas Kearns," 150; Rudd, "Samuel Newhouse," 300, 304; *Deseret Evening News*, 16 April 1918; and Bliss, *Merchants and Miners in Utah*, 229.

55. Thompson, *The Very Rich Book*, 194–95, 198; *Denver Republican*, 9 April 1910; *New York Times*, 13 September 1914; and Hall, *America's Successful Men of Affairs*, 1:444; 2:888.

56. Thompson, *The Very Rich Book*, 194.

57. High, "William Andrews Clark, Westerner," 254.

58. Hall, *America's Successful Men of Affairs*, 1:444, 742; 2:888; and Boner, *The Giant's Ladder*, 68–69.

59. See letter, Hammond to James Phelan, 13 August 1915; and the

Rocky Mountain Club membership list for 1915, Correspondence and Papers of James Phelan, 1880–1930, Bancroft Library, University of California, Berkeley. Keith's membership is noted in Stevens, ed., *Press Reference Library, Notables of the Southwest*, 111.

60. Hearst, *The Way It Was*, 37.

61. As quoted in Mack, "William Morris Stewart" 106. For more information on Stewart from the same source, see Effie M. Mack, "William M. Stewart" (Ph.D. diss., University of California, Berkeley, 1931).

62. Before election to the U.S. Senate, only Horace Tabor, John P. Jones, and Jerome Chaffee had held public office for more than three years. Of the rest, two had no previous experience in public office and most of the others had only one to two years. Several of the latter claimed service in state constitutional conventions as their political baptism. For the political careers of the ten senators, see Clifford P. Reynolds, ed., *Biographical Directory of the American Congress, 1774–1961* (Washington, D.C.: Government Printing Office, 1961). The brief political apprenticeship of the mining senators made them atypical. "Sixty-two percent of the membership in the 1870s were experienced politicians, having served over four years in national office or over six years in important state posts. By the 1890s, however, 78 percent of the Senate shared these qualifications. . . by 1900, the chamber was the preserve of professional politicians." See Rothman, *Politics and Power*, 128.

63. Larsen, "The Life of Thomas Kearns," 70.

64. Ibid., 86, 154.

65. Stewart, *Adolph Sutro*, 126–27; Angel, *History of Nevada*, 92, 97; Goodwin, *As I Remember Them*, 128–29. For an explanation of the business and other problems which allegedly prevented Sharon from participating in the Senate, see Bancroft, *Chronicles of the Builders of the Commonwealth*, 4:71–72.

66. Lewis, *Silver Kings*, 67; and Ostrander, *Nevada: The Great Rotten Borough*, 71–73.

67. Davis, ed., *The History of Nevada*, 1:422.

68. Ostrander, *Nevada: The Great Rotten Borough*, 72.

69. Lewis, *Silver Kings*, 168. For Fair's opinion of the senate, see Rothman, *Politics and Power*, 151.

70. As quoted in Lewis, *Silver Kings*, 167.

71. Johnson et al., eds., *Dictionary of American Biography*, 8:488.

72. Hearst, *The Way It Was*, 26.

73. Ibid., 26; Winfield J. Davis, *History of Political Conventions in California, 1849–1892* (Sacramento: California State Library, 1893), 436; Older, *George Hearst*, 195–96; and R. Hal Williams, *The Democratic Party and California Politics, 1880–1896* (Stanford: Stanford University Press, 1973), 102 n. 52.

74. *New York Times*, 19 January 1887. For information on Hearst's senatorial aspirations and alliance with Buckley, see Williams, *The Democratic Party and California Politics*, 25, 65–66, 105–7, 148–49; and William A. Bullough, *The Blind Boss and His City: Christopher Augustine Buckley and Nineteenth-Century San Francisco* (Berkeley: University of California Press, 1979), 160, 175–76, 183–84.

75. Smith, *Horace Tabor*, 207–26.

76. Tabor to McCourt, 20 February 188, folio file 14, no. 4, Tabor Papers.

77. Tabor to McCourt, 23 February 1883, folio file 15, no. 3, ibid.

78. Tabor to McCourt, 24 February 1883, folio file 15, no. 7, ibid.

79. Johnson et al., eds., *Dictionary of American Biography*, 3:590; *Denver Tribune-Republican*, 10 March 1886; *Portrait and Biographical Record of Denver and Vicinity, Colorado* (Chicago: Chapman Publishing Co., 1898), 129–30; and *History of the City of Denver, Arapahoe County, and Colorado* (Chicago: O. L Baskin and Co., 1880), 362–63.

80. Letter, Chaffee to Hill, 8 July 1878, as printed in unspecified newspaper in Dawson, Scrapbooks, Public Men, I, 61:3.

81. See also Smith, *Horace Tabor*, 94.

82. As quoted in Rothman, *Politics and Power*, 145. For Hill's senatorial career, see Bancroft, *Chronicles of the Builders of the Commonwealth*, 4:388–98.

83. For a complete analysis of the well-known Clark-Daly feud and Clark's belated senatorial election, see Malone, *The Battle for Butte*, 80–162. See also Ari Hoogenboom, "Industrialism and Political Leadership: A Case Study of the United States Senate," in Frederic C. Jaher, ed., *The Age of Industrialism in America: Essays in Social Structure and Cultural Values* (New York: The Free Press, 1968), 49–78.

84. Farrell, "William Andrews Clark," 62, 71.

85. For Clark's record on various political issues, see ibid., 62–70. See also Malone, *The Battle for Butte*, 195; Malone, "Midas of the West," 14; and High, "William Andrews Clark, Westerner," 260.

86. Farrell, "William Andrews Clark," 35 n. 1.

87. Malone, "Midas of the West," 14, and Malone, *The Battle for Butte*, 195.

88. As quoted in High, "William Andrews Clark, Westerner," 257.

89. Hammond, *Autobiography*, 2:483.

90. McLean, *Father Struck It Rich*, 55.

91. *Denver Times*, 2 August 1910.

92. See Hoyt, *The Guggenheims and the American Dream*, 181–86.

93. As quoted in Wells Drury, *An Editor on the Comstock Lode* (New York: Farrar and Rinehart, 1936), 270.

94. Angel, *History of Nevada*, 622.

95. G. T. Ingham, *Digging Gold among the Rockies* (Philadelphia: Hubbard Bros., 1880), 483.

96. Lavender, *Nothing Seemed Impossible*, 264.

97. *Oakland Tribune*, 23 April 1961.

98. Bliss, *Merchants and Miners in Utah*, 227.

99. Phelps, *Contemporary Biography of California's Representative Men*, 1:327.

100. McLean, *Father Struck It Rich*, 56.

101. Bancroft, *Chronicles of the Builders of the Commonwealth*, 4:243, 257–59.

102. Joaquin Miller, *An Illustrated History of the State of Montana* (Chicago: Lewis Publishing Co., 1894), 206–7.

103. Hubert H. Bancroft, "Colonel John D. Fry, a Biographical Sketch," ca. 1890, Bancroft Library, 5–6.

104. Barrett, "When Big Money Came to Owyhee," 25 n. 92.

105. Jennie Knight, "Biography of Amanda," box 168, p. 15, Knight Papers.

106. Whitney, *History of Utah*, 4:697.

107. Johnson et al., eds., *Dictionary of American Biography*, 10:189.

108. *Denver News*, 15 September 1902; Waters, *Midas of the Rockies*, 212–22.

109. For Amanda Reed's account of the journey, see Lyman, "Simeon Gannett Reed," 94–95.

110. Ibid., 95.

111. As quoted in Paul G. Merriam, "Urban Elite in the Far West: Portland, Oregon, 1870–1890," *Arizona and the West* 18 (Spring 1976): 47.

112. *Salt Lake Tribune*, 17 May 1901.

113. Horace Tabor, "Life of H. A. W. Tabor," 1889, Bancroft Library, 15. See financial statements for European lodging, folio file 522, nos. 1–8, Tabor Papers.

114. Letter, Nathaniel P. Hill to Alice Hill, 29 March 1866, Original Correspondence, folio 3, Hill Papers.

115. Letter, Nathaniel P. Hill to Alice Hill, 14 April 1866, ibid.

116. Hearst, *The Way It Was*, 38.

117. Letter, Phoebe Hearst to Jeannie Peck (Orrin Peck's mother), 15 April 1885, Orrin M. Peck Papers.

118. Thorndike, *The Very Rich*, 20–21; Simon, *Fifth Avenue*, 173–86. For Consuelo's marriage to and eventual divorce from the Duke, see her autobiography, Consuelo Vanderbilt Balsan, *The Glitter and the Gold* (New York: Harper and Bros., 1952).

119. *San Francisco Chronicle*, 5 August 1970.

120. *Portrait and Biographical Record of Denver and Vicinity, Colorado*, 130.

121. See *San Francisco Chronicle*, 4 June 1890; 5 April 1899; and Lewis, *Silver Kings*, 170–72, 180.

122. Bancroft, *Chronicles of the Builders of the Commonwealth*, 4:44, 78–79; and *Alta California*, 14 November 1885.

123. Helen M. Crampton, "A Sketch of the Kearns Family," MS, 1961, Utah State Historical Society, 3.

124. Lewis, *Silver Kings*, 102.

125. McLean, *Father Struck It Rich*, 110–20.

126. Letter, Vinson Walsh to his mother, 1905, container 1, file 1905, Walsh Papers.

127. Galbraith, *The Age of Uncertainty*, 68.

128. or Edward McLean's emotional problems and his wife's drug addiction, see McLean, *Father Struck It Rich*, 179–84, 292–93. The McLeans eventually divorced and Evalyn's former husband died in a mental institution.

129. See letter, George to Phoebe Hearst, 17 July (year undated), Incoming Correspondence, PAH Papers.

130. McLean, *Father Struck It Rich*, 52.

131. Smith, *Horace Tabor*, 228.

132. Ibid., 229; Lewis C. Gandy, *The Tabors: A Footnote of Western History* (New York: Press of the Pioneers, 1934), 251.

133. Rudd, "Samuel Newhouse," 304.

134. Tilton, *William Chapman Ralston*, 188.

135. *Rocky Mountain News*, 23 January 1881.

136. Lewis, *Silver Kings*, 51, 163.

137. Ibid., 88–89.

138. As quoted in Rudd, "Samuel Newhouse," 304.

139. Ibid., 304.

140. Knight, "Biography of Amanda," box 168, p. 15, Knight Papers.

141. Lester, *Brigham Street*, 96–97.

142. Stewart, *Adolph Sutro*, 32.

143. Hammond, *Autobiography*, 1:290.

144. Ibid., 2:761–72.

145. *National Cyclopedia of American Biography*, 24:387.

146. Ibid., 26:319.

147. Dana, *The Man Who Built San Francisco*, 222.

148. Swanberg, *Citizen Hearst*, 7–15.

149. Older, *William Randolph Hearst*, 29. A former editor of the *San Francisco Bulletin*, Fremont Older edited Hearst's *San Francisco Call* in the 1920s. After Hearst purchased the *Bulletin* in 1929, Older became the editor of the merged *Call-Bulletin* until his death in 1935.

150. Mangam, *The Clarks*, 54–55, 83; Malone, "Midas of the West," 8, 17.

151. Bancroft, "Colonel John D. Fry, a Biographical Sketch," 5.

152. Bancroft, *Chronicles of the Builders of the Commonwealth*, 4:44, 78–79.

153. Stewart, *Adolph Sutro*, 141; Hart, *A Companion to California*, 448; and Betty Magovern (Executive Assistant, Sutro and Co., Inc.), "Letter to the Editor," *California Historical Courier* 28 (June/July 1986): 2.

154. For a biography of Broadwater's son, see *National Cyclopedia of American Biography*, 45:175.

155. Letter, Thomas Walsh to his son, 20 July 1904, Family Correspondence, container 1, file 1900–1904, Walsh Papers.

156. McLean, *Father Struck It Rich*, 49–51, 57–58, 86.

157. Bancroft, *Chronicles of the Builders of the Commonwealth*, 4:397.

158. Mangam, *The Clarks*, 118, 123, 142; *Who Was Who in America, 1897–1942*, p. 225.

159. See *World Book Encyclopedia* (Chicago: World Book–Childcraft International, Inc., 1981), 9:33–34.

160. *National Cyclopedia of American Biography*, 19:288; Miller, *An Illustrated History of the State of Montana*, 803–4.

161. Jerome C. Smiley, *Semi-Centennial History of the State of Colorado* (Chicago: Lewis Publishing Co., 1913), 2:48.

162. Crampton, "A Sketch of the Kearns Family," 2, 4.

163. Knight, *The Jesse Knight Family*, 111.

164. Stewart, *Adolph Sutro*, 140–41.

165. Swanberg, *Citizen Hearst*, 22–23.

166. *Henry Newell*, 7.

167. High, "William Andrews Clark," 262.

168. Lewis, *Silver Kings*, 182–87; and *New York Herald*, 24 October 1893.

169. *San Francisco Chronicle*, 5 August 1970.

170. Lewis, *Silver Kings*, 111.

171. McLean, *Father Struck It Rich*, 122, 155–57, 171–79, 191.

172. Swanberg, *Citizen Hearst*, 22–23, 29–37, 52–53, 61–62.

173. Carol Dunlap, *California People* (Salt Lake City: Peregrine Smith Books, 1982), 89. Phoebe Hearst was sometimes very reluctant, if not actually opposed, to giving money to advance her son's newspaper career. Judging by a complaining letter to her close friend, Orrin Peck, in 1899, her son's driving ambition tested even her ample generosity: "Will is *insisting* upon buying a paper in *Chicago*. Says he will come over [to Paris] to see me if I do not go home very soon. It is *impossible* for me to throw away more money in any way for the simple reason that he has already absorbed almost all. . . . It is *madness*. I never know when or how W[m] will break out into some *additional* expensive scheme. . . no argument can induce me to commit such a folly as that of starting another newspaper." See Letter, Phoebe to Orrin Peck, 30 May 1899, Orrin M. Peck Papers.

174. Letter, Lloyd Tevis to Louise, 2 December 1897, Part II, box 5 Sharon Papers.

175. Letter, Tevis to Louise, 15 August 1893, ibid.

176. Letter, Knight to Raymond, 31 December 1908, Knight Investment Co. Correspondence, 1908–9, box 49, file I-J, Knight Papers.

177. As quoted in Mehls, "David H. Moffat, Jr.," 329.

178. Hammond, *Autobiography*, 1:209.

179. *San Francisco Examiner*, 26 July 1899.

5

THE FRONTIER THESIS AND THE LIFESTYLE OF WESTERN MINING ENTREPRENEURS

THE VAST MAJORITY OF THE MINING entrepreneurs in this study went west when mining was still a frontier industry, often joining the initial rushes to the various mining districts. Therefore, it seems appropriate to evaluate their lifestyle within the context of the familiar frontier thesis developed by Frederick Jackson Turner. As Turner pondered the 1890 announcement of the United States Bureau of the Census that the frontier had closed, he found reason to recall America's frontier heritage: "Not the constitution, but free land and an abundance of natural resources open to a fit people, made the democratic type of society in America. . . . These free lands promoted individualism, economic equality, freedom to rise, democracy."[1] Thus, individualism was one of many character traits allegedly produced by the frontier experience.

Judging by what the western mining leaders said about their success and by the evaluations of contemporary biographical encyclopedias, they regarded themselves as rugged individualists who succeeded by being industrious, ambitious, opportunistic, frugal, and honest. However, it is impossible to determine to what extent these self-proclaimed and sometimes exaggerated character traits were the product of the frontier experience or of the general conditioning of America's capitalist culture and Protestant religious heritage. Certainly, many leaders of eastern industries attributed their success to the same individualistic qualities mentioned by the western mining leaders.[2] Indeed, the general popularity of Horatio Alger's stories of poor, achieving youth; other varieties of so-called success or self-help literature, and the teachings of social Darwinism undoubtedly tended to standardize and nationalize the ideology of individualistic success in business.

Despite Turner's belief that individuals were more apt to make it on their own in the West than elsewhere in the nation, the mining moguls, like early pioneers, demanded aid from their government.[3] The western mining entrepreneurs and national industrial leaders alike welcomed government subsidy when it was to their advantage and protested government regulation when it threatened their interests. Some fought long and hard for the free and unlimited coinage of silver and a protective tariff on the importation of competing foreign lead, but resisted federal laws restricting the cutting of timber on public land or state initiatives to tax mines based on the estimated value of hidden ores. Although the mining leaders may have considered themselves rugged individualists eager for maximum freedom to operate their enterprises, they were quick to call for government aid when such freedom could not deliver the high profits they desired. Thus, their perception and practice of industrial success paradoxically combined the force of individual initiative with the safety net of government subsidy.

Their generous philanthropic support of education raises questions about the anti-intellectualism and cultural limitations identified historically with the West. Turner claimed that the pioneering experience emphasized materialistic attitudes over intellectual and aesthetic values, and modern scholars since have asserted that the frontier was a principal force contributing to American anti-intellectualism.[4] Ray Allen Billington suggests that "Americans today still reflect that background by placing the 'intellectual' well below the 'practical businessmen' in their ranking of folk heroes."[5] Nonetheless, the willingness of the mining magnates to support educational institutions indicates that the West, despite its initial lack of cultural refinement, did not entirely discount the value of academic learning and aesthetic appreciation.[6] Even practical business leaders could be intellectuals in their desire to improve the educational and artistic level of western society and to bring honor and prestige to themselves.

The apparent tendency of the western mining nouveaux riches to imitate the philanthropy and lifestyle of established eastern business elites also diminishes the cultural innovation and distinctiveness that Turner related to the frontier experience. In his classic analysis he asserted that "what the Mediterranean Sea was to the Greeks, breaking the bond of custom, offering new experiences, calling out new institutions and activities, that, and more, the ever retreating frontier has been to the United States. . . ."[7] However, once the mining entrepreneurs became successful, they did not seek to create a novel western lifestyle. Although they were less likely to live permanently in the East than commonly suspected, they looked east for models of proper elitist behavior. Many western moguls not only gave generously to charity, but built impressive mansions, threw lavish parties, patronized the arts and letters, purchased country estates and horse-racing stables, joined prestigious gentlemen's clubs, aspired to a seat in the U.S. Senate, took the grand tour of Europe, or indulged their families with expensive gifts and their children with exclusive educations. Eastern elites set the pattern, creating the social role that the western nouveaux riches of the Gilden Age fulfilled.[8] The social interaction of the mining magnates with their eastern counterparts in fraternal organizations and at society parties, in the enjoyment of common hobbies, and in the business and political arena not only led occasionally to well-publicized East-West marriages of their respective children, but undoubtedly helped to define the social expectations and obligations of business success for the impressionable Westerners. According to a student of western urban history, "in imposing modes of class distinction and norms of cultural excellence standardized in the East, the young western elites [of San Francisco and Denver] continued to transplant stereotypes of American society. Adapting eastern conventions to gain prestige, they curtailed indigenous growth."[9]

In sum, the western mining leaders were very much a part of the predominant business culture of America at the turn of the century. Indeed, their lifestyle illustrates the connection between the West and the rest of the nation. Far from being true pioneers in the Turnerian context, the mining elites perhaps sought to overcome a sense of provincial cultural inferiority by conforming to national standards of business elitism. Their resulting behavior exemplified cultural continuity rather than the cultural change emphasized by Turner's famous frontier thesis. Similarly, Turner's emphasis on the individualism and anti-intellectualism of the West should be qualified in view of the mining leaders' appeal for government aid on the one hand and their willingness to support education and patronize the arts and letters on the other. In short, the behavior of the bonanza kings reflects the homogeneity of national culture more than a separate regional identity.

NOTES

1. Frederick Jackson Turner, *The Frontier in American History* (New York: Henry Holt and Co., 1920), 259, 293. The first part of this quotation originally appeared in a commencement address at the University of Washington on 17 June 1914.

2. See Irvin G. Wyllie, *The Self-Made Man in America: The Myth of Rags to Riches* (New York: The Free Press, 1966), 34–54.

3. For an effective challenge to Turner's argument on behalf of frontier-bred individualism, see Moody C. Boatright, "The Myth of Frontier Individualism," *Southwestern Social Science Quarterly* 22 (June 1941): 14–32.

4. See Ray Allen Billington, *America's Frontier Heritage* (New York: Holt, Rinehart and Winston, 1966), 95; and Merle Curti, "Intellectuals and Other People," *American Historical Review* 60 (January 1955): 259–82.

5. Ray Allen Billington, *The American Frontier Thesis: Attack and Defense* (Washington, D.C.: American Historical Association, 1971), 41.

6. For evidence that Westerners maintained cultural ties with the East, see Louis B. Wright, "The Westward Advance of the Atlantic Frontier," *Huntington Library Quarterly* 11 (May 1948): 261–75, and *Culture on the Moving Frontier* (Bloomington: Indiana University Press, 1955; New York: Harper and Brothers, 1961); Earl Pomeroy, "Toward a Reorientation of Western History: Continuity and Environment," *Mississippi Valley Historical Review* 41 (March 1955): 582–83, 591–92; and Ronald L. Davis, "Culture on the Frontier," *Southwest Review* 53 (Autumn 1968): 383–403.

7. Frederick Jackson Turner, "The Significance of the Frontier in American History," *American Historical Association Annual Report, 1893* (Washington, D.C., 1894), 227.

8. The behavior of elites varied in different cities. "In their quest for notoriety, as in other characteristics such as lavish lifestyles, great wealth, and open entry, the New Yorkers more closely resembled the elites of new cities like Chicago, Denver, and San Francisco than did the mellower Atlantic seaboard aristocrats" (Frederick Cople Jaher, "Style and Status: High Society in Late Nineteenth-Century New York," in Jaher, ed., *The Rich, the Well Born, and the Powerful: Elites and Upper Classes in History* [Urbana: University of Illinois Press, 1973, 281). For evidence that western elites copied the lifestyle of eastern elites, see Merriam, "Urban Elite in the Far West," 41–52. On the same theme, see Gunther Barth, "Metropolism and Urban Elites in the Far West," in Jaher, ed., *The Age of Industrialism in America*, 158–87.

9. Barth, "Metropolism and Urban Elites in the Far West," 179.

Appendix

Western Mining Leaders, 1870–1900

Lemuel "Sandy" Bowers	1833–1868	California, Nevada*
Charles A. Broadwater	1840–1892	Montana
Bela S. Buell	1836–1918	Colorado
Jerome Chaffee	1825–1886	Colorado
William A. Clark	1839–1925	Montana, Arizona
Marcus Daly	1841–1900	Montana
Andrew J. Davis	1819–1890	Montana
Joseph R. DeLamar	1843–1918	Idaho, Utah
James G. Fair	1831–1894	Nevada
James C. Flood	1826–1889	Nevada
John D. Fry	1819–1900	Nevada
James B. Grant	1848–1911	Colorado
James Ben Ali Haggin	1822–1914	South Dakota
John Hays Hammond	1855–1936	Idaho

*Areas of primary activity. Most of these individuals invested in mining areas throughout the West and even in Latin America.

Albert K. P. Harmon	1821–1896	Nevada
Samuel T. Hauser	1833–1914	Montana, Idaho
Alvinza Hayward	1821–1904	California, Nevada
George Hearst	1820–1891	South Dakota
F. Augustus Heinze	1869–1914	Montana
Nathaniel P. Hill	1832–1900	Colorado
Irving Howbert	1846–1934	Colorado
John P. Jones	1829–1912	California, Nevada
Thomas Kearns	1862–1918	Utah
David Keith	1847–1918	Utah
Jesse Knight	1845–1921	Utah
John W. Mackay	1831–1902	Nevada
Darius O. Mills	1825–1910	Nevada, Idaho
David H. Moffat	1839–1911	Colorado
Henry Newell	1844–1928	Utah
Samuel Newhouse	1853–1930	Utah, Colorado
William S. O'Brien	1826–1878	Nevada
John Q. Packard	1822–1908	Utah
Almarin B. Paul	1823–1909	California
Thomas C. Power	1939–1923	Montana
William C. Ralston	1826–1875	Nevada
Simeon G. Reed	1830–1895	Idaho
Thomas Selby	1820–1875	California
William C. Sharon	1821–1885	Nevada
Dennis Sheedy	1846–1923	Colorado
Winfield S. Stratton	1848–1902	Colorado
Adolph H. Sutro	1830–1898	Nevada
Horace A. Tabor	1830–1899	Colorado
Lloyd Tevis	1824–1899	South Dakota
Joseph "Rob" Walker	1836–1901	Utah
Matthew H. Walker	1845–1916	Utah
Enos Wall	1839–1920	Utah
Thomas F. Walsh	1851–1910	Colorado

Simeon Wenban	1824–?	Nevada
Jerome B. Wheeler	1841–1918	Colorado
Henry R. Wolcott	1846–1921	Colorado

Bibliography

MANUSCRIPTS

Brigham Young University Archives, Provo, Utah

Business Records and Family Papers of Jesse Knight.

California State Library, Sutro Branch, San Francisco, California

Archibald C. Unsworth "Memoir." Material on Adolph Sutro. Vol. 1. Unsworth was Sutro's traveling companion and secretary.

Colorado College, Colorado Springs, Colorado

Fagan, George V. (Head Librarian). Letter, August 5, 1982, to the author regarding Irving Howbert.

Henry E. Huntington Library and Art Gallery, San Marino, California

Orrin M. Peck Papers. Peck was a close friend of both Phoebe and William Randolph Hearst.

Library of Congress, Manuscript Division, Washington, D.C.

Evalyn Walsh McLean Family Papers, 1886–1948. Evalyn Walsh McLean was the daughter of Thomas Walsh. The papers contain his personal and business correspondence, 1887–1903.

Reed College, Portland, Oregon

Reed College Project, Division of Professional and Service Projects, W.P.A., 1940. Letters and Private Papers of Simeon G. Reed. 40 volumes.

San Mateo County Historical Association, San Mateo, California

Bromfield, Davenport. "Alvinza Hayward, Pioneer," n.d.

State Historical Society of Colorado, Denver, Colorado

Horace A. W. Tabor Family Papers

Nathaniel P. Hill Papers

Thomas F. Dawson, comp. Scrapbooks. Eighty volumes of clippings from Colorado newspapers collected by the former director of the Historical Society.

Spring, Agnes W., comp. "Notes and Extracts from the Personal Papers of David H. Moffat and Jerome B. Chaffee." The original, complete collection of the Moffat-Chaffee Papers is in the First National Bank, Denver.

University of California, Bancroft Library, Berkeley, California

Bancroft, Hubert H. "Biographical Sketch," in Dictation and Biographical Material on Simeon G. Reed, 1888.

———. "Biographical Sketches of Thomas Henry Selby," ca. 1890.

———. "Biography of Darius O. Mills," ca. 1885.

———. "Colonel John D. Fry, a Biographical Sketch," ca. 1890.

Correspondence and Papers of James Phelan, 1880–1930.

Henry Walcott. "Dictation and Related Biographical Material for David H. Moffat," 1885; and (by Walter Cheesman), 1885.

George Holmes Howison Papers, 1862–1917.

Phoebe Apperson Hearst Papers.
Sandy Bowers, material in "California Biography," undated.
Tabor, Horace A. W. "Life of Horace Tabor" (autobiography), 1889.
Victor, Frances F. "Biographical Sketch of John C. Ainsworth," in Dictation and Biographical Material on John C. Ainsworth, 1890.
Wenban, Simeon. "Dictations and Biographical Material," 1888.
William Sharon Family Papers.
Wolcott, Henry R. "Dictation," 1888.

University of California, Special Collections Department, Los Angeles, California
John P. Jones Family Papers.

University of Utah, Marriott Library, Salt Lake City, Utah
The Alta Club Papers, 1883–1984.

Utah State Historical Society, Salt Lake City, Utah
Crampton, Helen M. "A Sketch of the Kearns Family," 1961.

NEWSPAPERS

Various newspapers provided obituaries or interviews of the mining leaders as well as other kinds of editorial comment and essential information:

Alta California (San Francisco). 8 September 1869; 15 January 1871; 10 June 1875; 22 October 1878; 1 January 1880; 1, 10 January 1881; 2 October, 14 November 1884.
Colorado Springs Gazette. 11, 12 July, 16 August 1899.
Colorado Transcript (Golden). 6 February 1901.
Denver News. 15 September 1902; 1 June 1921.
Denver Polly Pry. Ca. 1914.

Denver Post. 1904; 10 April 1910; 19 March 1911; 1, 24 June 1921; 25 February 1932; 21 December 1934.

Denver Republican. 7 June 1904; 9 April 1910.

Denver Times. 27 January 1904; 2 August 1910.

Deseret Evening News (Salt Lake City). 20 March 1907; 16 April 1918.

Helena Weekly Herald. 16 April 1885.

Holiday Miner (Butte). 1 January 1888.

Los Angeles Herald-Examiner. 26 September 1976.

Mining and Scientific Press (San Francisco). 10 July 1920.

New York Herald. 24 October 1883.

New York Herald Tribune. 3 March 1925.

New York Times. 19 January 1887; 25 May 1892; 28 May 1904; 3, 13 September, 5 November 1914; 2 December 1918; 18 April 1919; 26 August 1923; 3 March 1925.

Oakland Enquirer. 30 April, 8 May 1896.

Oakland Tribune. 23 April 1961; 27 April 1969.

Rocky Mountain News (Denver). 20 July 1879; 23 January, 28 August, 20 November 1881; 1 July 1883; 9 April 1910.

Salt Lake Tribune. 7 July 1888.

San Francisco Call. 11 January 1880; 4 November 1883.

San Francisco Chronicle. 4 June 1890; 27 March 1966; 5, 6, 20 August 1970.

San Francisco Examiner. 26 July 1899; 14–16 April 1919.

San Francisco Examiner and Chronicle. 14, 21 August 1966.

Territorial Enterprise (Virginia City, Nevada). 6 May 1892.

Tribune-Republican (Denver). 24 January 1886.

CONTEMPORARY ARTICLES, BOOKS, DOCUMENTS, AND REPORTS

"American Millionaires and Their Public Gifts." *Review of Reviews* 7 (February 1893): 48–60.

Angel, Myron. *History of Nevada with Illustrations and Biographical Sketches of Its Prominent Men and Pioneers*. Berkeley: Howell-North, 1948 (orig. pub. 1881).

Bancroft, Hubert H. *Chronicles of the Builders of the Commonwealth*. 7 vols. San Francisco: The History Co., 1891–92.

Bryce, James. *The American Commonwealth*. 3d ed. 2 vols. London: Macmillan and Co., 1893.

Carnegie, Andrew. "The Best Fields for Philanthropy." *North American Review* 149 (1889): 687–96.

———. "Wealth." *North American Review* 148 (1889): 657–62.

Davis, Winfield J. *History of Political Conventions in California, 1849–1892*. Sacramento: California State Library, 1893.

Elite Directory of 1879. San Francisco: The Argonaut Publishing Co., 1879.

Godkin, E. L. "The Expenditure of Rich Men." *Scribner's Magazine* 20 (1896): 495–501.

Hall, Henry, ed. *America's Successful Men of Affairs: An Encyclopedia of Contemporaneous Biography*. 2 vols. New York: New York Tribune, 1895–96.

Hill, Nathaniel P. "Address delivered at Leadville, Colorado, before the YMCA, April 12, 1888," in *Speeches and Papers on the Silver, Postal Telegraph, and Other Economic Questions*. Colorado Springs: Gazette Printing Co., 1890.

History of the City of Denver, Arapahoe County, and Colorado. Chicago: O. L. Baskin and Co., 1880.

Ingham, G. T. *Digging Gold among the Rockies*. Philadelphia: Hubbard Bros., 1880.

McAllister, Samuel Ward. *Society As I Have Found It*. New York: Cassell, 1890.

Miller, Joaquin. *An Illustrated History of the State of Montana*. Chicago: Lewis Publishing Co., 1894.

Phelps, Alonzo. *Contemporary Biography of California's Representative Men*. 2 vols. San Francisco: A. L. Bancroft and Co., 1881.

Portrait and Biographical Record of Denver and Vicinity, Colorado. Chicago: Chapman Publishing Co., 1898.

Reed College Record (Portland), May 1911.

Scott, Harvey, ed. *History of Portland, Oregon, with Illustrations and Biographical Sketches of Prominent Citizens and Pioneers.* Syracuse: D. Mason and Co., 1890.

Swasey, W. F. *The Early Days and Men of California.* Oakland: Pacific Press Publishing Co., 1891.

Teetor, Henry D. "Honorable H. A. W. Tabor, Ex-United States Senator, and Ex–Lieutenant-Governor of Colorado." *Magazine of Western History* 9 (January 1889): 268–73.

U.S. Congress. Senate. U.S. Monetary Commission. *Report and Accompanying Documents*, Senate Report 703. 2 vols. 44th Congress, 2d session. Washington, D.C.: Government Printing Office, 1877–79.

University of California Regents. *Annual Report of the Secretary of the Board of Regents* [for the year ending June 30, 1896]. Sacramento: State Printing Office, 1896.

Whitney, Orson F. *History of Utah.* 4 vols. Salt Lake City: George Q. Cannon and Sons, 1892–1904.

PUBLISHED MEMOIRS AND REMINISCENT ACCOUNTS

Carr, Clark E. *My Day and Generation.* Chicago: A. C. McClurg and Co., 1908.

Dawson, Thomas F. *Life and Character of Edward Oliver Wolcott, Late a Senator of the United States from the State of Colorado.* 2 vols. New York: Knickerbocker Press, 1911.

Drury, Wells. *An Editor on the Comstock Lode.* New York: Farrar and Rinehart, 1936.

Goodwin, Charles C. *As I Remember Them.* Salt Lake City: Salt Lake Commercial Club, 1913.

Gorham, Harry M. *My Memories of the Comstock.* Los Angeles: Suttonhouse Publishers, 1939.

Hammond, John Hays. *The Autobiography of John Hays Hammond.* 2 vols. New York: Farrar and Rinehart, 1935.

Hearst, George. *The Way It Was: Recollections of U.S. Senator George Hearst, 1820–1891.* San Francisco: The Hearst Corporation, 1972.

Howbert, Irving. *Memories of a Lifetime in the Pikes Peak Region.* New York: G. P. Putnam's Sons, 1925.

McLean, Evalyn Walsh. *Father Struck It Rich.* Boston: Little, Brown and Co., 1936.

Reed, Simeon G. *Letters and Private Papers of Simeon G. Reed.* 40 vols. Portland, Ore.: Reed College Project, Works Progress Administration, 1940. *See also* manuscripts section, Reed, Simeon G.

Sheedy, Dennis. *The Autobiography of Dennis Sheedy.* Denver: privately printed, 1922.

THESES AND DISSERTATIONS

Farrell, Mary M. "William Andrews Clark." M.A. thesis, University of Washington, 1933.

Hakola, John W. "Samuel T. Hauser and the Economic Development of Montana: A Case Study in Nineteenth-Century Frontier Capitalism." Ph.D. dissertation, Indiana University, 1961.

Larsen, Kent S. "The Life of Thomas Kearns." M.A. thesis, University of Utah, 1964.

Lyman, Lloyd Gouvy. "Simeon Gannett Reed, A Preliminary Biographical Study." B.A. thesis, Reed College, 1948.

Mack, Effie M. "William M. Stewart." Ph.D. dissertation, University of California, Berkeley, 1931.

McNelis, Sarah Ann. "The Life of F. Augustus Heinze." M.A. thesis, Montana State University, 1947.

Mehls, Steven F. "David H. Moffat, Jr.: Early Colorado Business

Leader." Ph.D. dissertation, University of Colorado, Boulder, 1982.

Myers, James E. "The Educational Work of Andrew Jackson Moulder." Ph.D. dissertation, University of California, Berkeley, 1961.

Reese, Gary Fuller. "'Uncle Jesse,' The Story of Jesse Knight, Miner, Industrialist, Philanthropist." M.S. thesis, Brigham Young University, 1961.

Roland, Carol. "The California Kindergarten Movement: A Study in Class and Social Feminism." Ph.D. dissertation, University of California, Riverside, 1980.

Smith, Keith L. "A History of the Brigham Young University—The Early Years, 1875–1921." Ph.D. dissertation, Brigham Young University, 1972.

SECONDARY ARTICLES AND BOOKS

Abeloe, William N., et al. *Historic Spots in California*. 3d ed. Stanford: Stanford University Press, 1966.

Akers, Dwight. "David H. Moffat and His Home Town." *Colorado Magazine* 27 (July 1950): 212–14.

Arrington, Leonard J., and Gary B. Hansen. *The Richest Hole on Earth: A History of the Bingham Copper Mine*. Logan: Utah State University Press, 1963.

Atherton, Lewis. *The Cattle Kings*. Bloomington: Indiana University Press, 1961.

Bakken, Gordon M. "The Taxation of Mineral Wealth and the Nevada Constitutional Convention of 1864." *Nevada Historical Society Quarterly* 12 (Winter 1969): 5–15.

Balsan, Consuelo Vanderbilt. *The Glitter and the Gold*. New York: Harper and Bros., 1952.

Bannister, Robert C. *Social Darwinism: Science and Myth in Anglo-American Social Thought*. Philadelphia: Temple University Press, 1979.

Barrett, G. W. "Colonel E. A. Wall: Mines, Miners, and Mormons." *Idaho Yesterdays* 14 (Fall 1970): 3–11.

———. "Enos Andrew Wall, Mine Superintendent and Inventor." *Idaho Yesterdays* 15 (Spring 1971): 24–31.

———. "When Big Money Came to Owyhee." *Idaho Yesterdays* 13 (Spring 1969): 2–9, 22–29.

Barth, Gunther. "Metropolism and Urban Elites in the Far West." In *The Age of Industrialism in America: Essays in Social Structure and Cultural Values*, edited by Frederic C. Jaher, 158–87. New York: The Free Press, 1968.

Bayne, Nedra. "The Broadwater: Relic of Elegance." *Montana: The Magazine of Western History* 19 (Summer 1969): 58–66.

Beebe, Lucius. *The Big Spenders*. New York: Simon and Schuster, 1967.

Bergsma, Donald J. "Surprising Structure behind a Classic Facade." *Western Architect and Engineer* 219 (May 1960): 32–33.

Betts, John R. *America's Sporting Heritage: 1850–1950*. Reading, Mass.: Addison-Wesley Publishing Co., 1974.

Billington, Ray Allen. *The American Frontier Thesis: Attack and Defense*. Washington, D.C.: American Historical Association, 1971.

———. *America's Frontier Heritage*. New York: Holt, Rinehart and Winston, 1966.

Biographical Record of Salt Lake City and Vicinity. Chicago: National Historical Record Co., 1902.

Black, David. *The King of Fifth Avenue: The Fortunes of August Belmont*. New York: Dial Press, 1981.

Bliss, Jonathan. *Merchants and Miners in Utah: The Walker Brothers and Their Bank*. Salt Lake City: Western Epics, 1983.

Boatright, Moody C. "The Myth of Frontier Individualism." *Southwestern Social Science Quarterly* 22 (June 1941): 14–32.

Boner, Harold A. *The Giant's Ladder: David H. Moffat and His Railroad*. Milwaukee: Kalmbach Publishing Co., 1962.

Bonfils, Winifred B. *The Life and Personality of Phoebe Apperson Hearst*. San Francisco: John Henry Nash, 1928.

Brechin, Gray. "A Little Place in the Country." *California Monthly* 90 (October 1979): 12–14.

Bridges, Hal. "The Robber Baron Concept in American History." *Business History Review* 33 (Spring 1958): 1–13.

Bullough, William A. *The Blind Boss and His City: Christopher Augustine Buckley and Nineteenth-Century San Francisco*. Berkeley: University of California Press, 1979.

Byers, William N. *Encyclopedia of Biography of Colorado*. Chicago: Century Publishing and Engraving Co., 1901.

Cash, Joseph H. *Working the Homestake*. Ames: Iowa State University Press, 1973.

Chalmers, David. "From Robber Barons to Industrial Statesmen: Standard Oil and the Business Historians." *American Journal of Economics and Sociology* 20 (October 1960): 47–58.

Chambers, S. Allen, ed. *Discovering Historic America: California and the West*. New York: E. P. Dutton and Co., 1982.

Clanton, Gene. "'Hayseed Socialism' on the Hill: Congressional Populism, 1891–1895." *Western Historical Quarterly* 15 (April 1984): 1390–62.

———. "Silver Issue." In *The Reader's Encyclopedia of the American West*, edited by Howard R. Lamar, 1110–12. New York: Thomas Y. Crowell Co., 1977.

Clary, Raymond H. *The Making of Golden Gate Park: The Early Years, 1865–1906*. San Francisco: San Francisco Examiner, 1980.

Clinch, Thomas A. *Urban Populism and Free Silver in Montana*. Missoula: University of Montana Press, 1970.

Cochran, Thomas C. *Business in American Life: A History*. New York: McGraw-Hill, Inc., 1974.

———. "The Legend of the Robber Barons." *Pennsylvania Magazine of History and Biography* 76 (July 1950): 307–21.

———, and William Miller. *The Age of Enterprise: A Social History of Industrial America*. New York: Harper and Row, 1961.

Curti, Merle. "Intellectuals and Other People." *American Historical Review* 60 (January 1955): 259–82.

———, Judith Green, and Roderick Nash. "Anatomy of Giving: Millionaires in the Late 19th Century." *American Quarterly* 15 (1963): 416–35.

———, and Roderick Nash. *Philanthropy in the Shaping of American Higher Education*. New Brunswick: Rutgers University Press, 1965.

Dallas, Sandra. *Cherry Creek Gothic: Victorian Architecture in Denver*. Norman: University of Oklahoma Press, 1971.

Dana, Julian. *The Man Who Built San Francisco: A Study of Ralston's Journey with Banners*. New York: Macmillan Co., 1936.

Davis, Ronald L. Culture on the Frontier." *Southwest Review* 53 (Autumn 1968): 383–403.

Davis, Samuel P., ed. *The History of Nevada*. 2 vols. Reno: Elms Publishing Co., 1913.

Degler, Carl N., et al. *The Democratic Experience: An American History*. 5th ed. 2 vols. Glenview, Ill.: Scott, Foresman and Co., 1981.

Delgado, James. "Sutro Heights Park: Traces of a Dream." *American West* 18 (May/June 1981): 60–66.

Dillon, Richard H. *The Anatomy of a Library*. Sacramento: California State Printing Office, 1957.

Dorsett, Lyle W. *The Queen City: A History of Denver*. Boulder: Pruett Publishing Co., 1977.

Dubofsky, Melvyn. "The Origins of Western Working Class Radicalism, 1890–1905." *Labor History* 7 (Spring 1966): 131–54.

Dulles, Foster R. *A History of Recreation: America Learns to Play*.

New York: Appleton-Century-Crofts, 1965.

Dunlap, Carol. *California People.* Salt Lake City: Peregrine Smith Books, 1982.

Elliott, Russell R. *Servant of Power: A Political Biography of Senator William M. Stewart.* Reno: University of Nevada Press, 1983.

Encyclopedia of American History. Guilford, Conn.: The Dushkin Publishing Group, Inc., 1973.

Fahey, John. *The Days of the Hercules.* Moscow: University of Idaho Press, 1978.

Fell, James E., Jr. "Nathaniel P. Hill, A Scientist-Entrepreneur in Colorado." *Arizona and the West* 15 (Winter 1973): 315–32.

Ferrier, William W. *Ninety Years of Education in California, 1846–1936: A Presentation of Educational Movements and Their Outcome in Education Today.* Berkeley: Sather Gate Book Shop, 1937.

———. *Origin and Development of the University of California.* Berkeley: Sather Gate Book Shop, 1930.

Ferrill, Will, ed. *Sketches of Colorado.* 2 vols. Denver: Western Press Bureau Co., 1911.

Galbraith, John K. *The Age of Uncertainty.* Boston: Houghton Mifflin Co., 1977.

Gandy, Lewis C. *The Tabors: A Footnote of Western History.* New York: Press of the Pioneers, 1934.

"The Gilded Age of Butte, Still in Evidence." *Sunset* 171 (July 1983): 30–34.

Gille, Marcia. "Lady Luck and the Peepstone Palace." *True West* 31 (January 1985): 42–46.

Glass, Mary E. *Silver and Politics in Nevada, 1892–1902.* Reno: University of Nevada Press, 1969.

Glasscock, Carl B. *Lucky Baldwin, The Story of an Unconventional Success.* Indianapolis: Bobbs-Merrill Co., 1933.

———. *The War of the Copper Kings, the Builders of Butte and the*

Wolves of Wall Street. New York: Grosset and Dunlap, 1966 (orig. pub. 1935).

Godkin, E. L. "The Expenditure of Rich Men." In *The Nation Transformed, The Creation of an Industrial Society*, edited by Sigmund Diamond, 119–26. New York: George Braziller, 1963.

Gould, Milton S. *A Cast of Hawks*. La Jolla, Calif.: Copley Books, 1985.

Gregory, Frances W., and Irene D. Neu. "The American Industrial Elite in the 1870s, Their Social Origins." In *Men in Business: Essays in the History of Entrepreneurship*, edited by William Miller, 193–211. Cambridge, Mass.: Harvard University Press, 1952.

Guinn, James M. *History of the State of California and Biographical Record of the Sacramento Valley, California*. Chicago: Chapman Publishing Co., 1906 (orig. pub. 1902).

Hagar, George J. "Magnitude of American Benefactions." *Review of Reviews* 29 (April 1904): 464–65.

Hart, James D. *A Companion to California*. New York: Oxford University Press, 1978.

Henry Newell: An Account of the Events in His Life Leading to the Establishment of the Henry Newell Scholarships at Stanford University. Stanford: Stanford University Press, 1954.

Hicks, John D. *The Populist Revolt: A History of the Farmers' Alliance and People's Party*. Lincoln: University of Nebraska Press, 1961 (orig. pub. 1931).

High, James. "William Andrews Clark, Westerner: An Interpretive Vignette." *Arizona and the West* 2 (Autumn 1960): 245–64.

Historical Records Survey. *Guide to Depositories of Manuscript Collections in the United States: California*. Los Angeles: Southern California Historical Records Survey Project, 1941.

Hoag, Charles C., ed. *The San Francisco Blue Book*. San Francisco: Charles C. Hoag, 1902.

Hofstadter, Richard. *Social Darwinism in American Thought*. Rev. ed. Boston: Beacon Press, 1955.

———, and Walter P. Metzger. *The Development of Academic Freedom in the United States*. New York: Columbia University Press, 1955.

Holbrook, Stewart H. *The Age of the Moguls*. Garden City, N.Y.: Doubleday and Co., 1954.

Hoogenboom, Ari. "Industrialism and Political Leadership: A Case Study of the United States Senate." In *The Age of Industrialism in America: Essays in Social Structure and Cultural Values*, edited by Frederic C. Jaher, 49–78. New York: The Free Press, 1968.

Hornbein, Marjorie. "Davis Waite, Silver, and Populism." In *Essays and Monographs in Colorado History*, no. 1, edited by David N. Wetzel, 71–90. Denver: Colorado State Historical Society, 1983.

Hoyt, Edwin P., Jr. *The Guggenheims and the American Dream*. New York: Funk and Wagnalls, 1967.

Irvine, Leigh H., ed. *A History of the New California, Its Resources and People*. 2 vols. New York: Lewis Publishing Co., 1905.

Jaher, Frederic C. "Style and Status: High Society in Late Nineteenth-Century New York." In *The Rich, the Well Born, and the Powerful: Elites and Upper Classes in History*, edited by Frederic C. Jaher, 258–84. Urbana: University of Illinois Press, 1973.

James, Edward T., et al., eds. *Notable American Women, 1607–1950*. 3 vols. Cambridge, Mass: Harvard University Press, 1971.

Jensen, Vernon H. *Heritage of Conflict: Labor Relations in the Non-Ferrous Metals Industry up to 1930*. Ithaca: Cornell University Press, 1950.

Johnson, Allen, et al., eds. *Dictionary of American Biography*. 22 vols. and supplements. New York: Charles Scribner's Sons, 1928–1958.

Johnson, Claudius O. "The Story of Silver Politics in Idaho, 1892–1902." *Pacific Northwest Quarterly* 33 (July 1942): 283–96.

Josephson, Matthew. *The Robber Barons: The Great American Capitalists, 1861–1901*. New York: Harcourt, Brace and World, 1962 (orig. pub. 1934).

Kelsey, Harry E., Jr. *Frontier Capitalist: The Life of John Evans*. Denver: State Historical Society of Colorado, 1969.

Kirker, Harold. *California's Architectural Frontier: Style and Tradition in the Nineteenth Century*. Salt Lake City: Peregrine Smith, Inc., 1973.

Kirkland, Edward C. *Dream and Thought in the Business Community, 1860–1900*. Chicago: Quadrangle Books, 1964.

———. "The Robber Barons Revisited." *American Historical Review* 66 (October 1960): 68–73.

Kistler, W. H., Stationery Co. *Who's Who in Denver Society, 1908*. Denver: W. H. Kistler Stationery Co., 1908.

Knight, Jesse William. *The Jesse Knight Family, His Forebears and Family*. Salt Lake City: Deseret News Press, 1940.

Kohl, Edith E. *Denver's Historic Mansions: Citadels to the Empire Builders*. Denver: Sage Books, 1957.

Larson, Robert W. "Populism in the Mountain West: A Mainstream Movement." *Western Historical Quarterly* 13 (April 1982): 143–64.

Lavender, David. *Nothing Seemed Impossible: William C. Ralston and Early San Francisco*. Palo Alto: American West Publishing Co., 1975.

Lawrence, William. "Relation of Wealth to Morals." *World's Work* 1 (1901): 286–92.

Lester, Margaret D. *Brigham Street*. Salt Lake City: Utah State Historical Society, 1979.

Lewis, Oscar. *The Big Four: The Story of Huntington, Stanford, Hopkins, and Crocker, and of the Building of the Central Pacific*. New York: Alfred A. Knopf, 1938.

———. *Silver Kings: The Lives and Times of Mackay, Fair, Flood, and O'Brien, Lords of the Nevada Comstock Lode*. New York: Alfred A. Knopf, 1967 (orig. pub. 1947).

Littell, Robert. "The House that Clark Built." *New Republic* 30 (30 March 1927): 171–72.

Livesay, Harold C. *Andrew Carnegie and the Rise of Big Business*. Boston: Little, Brown and Co., 1975.

Lockwood, Charles. "Rincon Hill Was San Francisco's Most Genteel Neighborhood." *California History* 58 (Spring 1979): 49–61.

———. *Suddenly San Francisco: The Early Years of an Instant City*. San Francisco: The San Francisco Examiner, 1978.

McGiffert, Michael. *The Higher Learning in Colorado: An Historical Study, 1860–1940*. Denver: Sage Books, 1964.

McMechen, Edgar C. *The Moffat Tunnel of Colorado, An Epic of Empire*. 2 vols. Denver: Wahlgreen Publishing Co., 1927.

Mack, Effie Mona. "William Morris Stewart, 1827–1909." *Nevada Historical Society Quarterly* 7 (1964): 5–121.

Magovern, Betty. "Letter to the Editor." *California Historical Courier* 28 (June/July 1986): 2.

Malmquist, Orvin N. *The Alta Club, 1883-1974*. Salt Lake City: privately printed, 1974.

Malone, Michael P. *The Battle for Butte: Mining and Politics on the Northern Frontier, 1864–1906*. Seattle: University of Washington Press, 1981.

———. "Midas of the West: The Incredible Career of William Andrews Clark." *Montana: The Magazine of Western History* 33 (Autumn 1983): 2–17.

Mangam, William D. *The Clarks, An American Phenomenon*. New York: Silver Bow Press, 1941.

Manter, Ethel. *Rocket of the Comstock: The Story of John William Mackay*. Caldwell, Idaho: Caxton Printers, 1950.

May, Henry F. *Protestant Churches and Industrial America*. New York: Octagon Books, Inc., 1963 (orig. pub. 1949).

Merriam, Paul G. "Urban Elite in the Far West: Portland, Oregon, 1870–1890." *Arizona and the West* 18 (Spring 1976): 41–52.

Miller, Harriet, and Elizabeth Harrison. "Marcus Daly's Western Champions." *The Western Horseman* (March 1958): 26–27.

Mills, C. Wright. "The American Business Elite: A Collective Portrait." *Journal of Economic History* (special issue *The Tasks of Economic History*) 5 (December 1945): 20–44.

Moffett, Samuel E. "Darius Ogden Mills." *The Cosmopolitan* 33 (May–October 1902): 289–92.

———. "James Ben Ali Haggin." *The Cosmopolitan* 33 (May–October 1902): 166–67.

Morris, John R. *Davis H. Waite: The Ideology of a Western Populist.* Lanham, Md.: University Press of America, 1982.

Morris, Richard B., ed. *Encyclopedia of American History.* New York: Harper and Row, 1970.

Mrozek, Donald J. *Sport and American Mentality, 1880–1910.* Knoxville: University of Tennessee Press, 1983.

Muscatine, Doris. *Old San Francisco: The Biography of a City from Early Days to the Earthquake.* New York: G. P. Putnam's Sons, 1975.

Myers, Gustavus. *History of the Great American Fortunes.* 2 vols. Chicago: Charles H. Kerr and Co., 1910.

National Cyclopedia of American Biography. 51 vols. New York: James T. White and Co., 1892–1969.

Nevins, Allan. *Study in Power: John D. Rockefeller, Industrialist and Philanthropist.* 2 vols. New York: Charles Scribner's Sons, 1953.

Nunis, Doyce B., Jr. "Kate Douglas Wiggin: Pioneer in California Kindergarten Education." *California Historical Society Quarterly* 41 (December 1962): 291–307.

Older, Cora. *William Randolph Hearst, American.* New York: D. Appleton-Century Co., 1936.

Older, Fremont and Cora Older. *George Hearst, California Pioneer.*

Los Angeles: Westernlore Press, 1966 (orig. pub. 1933).

Ostrander, Gilman M. *Nevada, the Great Rotten Borough, 1859–1964*. New York: Alfred A. Knopf, 1966.

Paine, Swift. *Eilley Orrum, Queen of the Comstock*. Palo Alto: Pacific Books, 1950.

Paul, Rodman W. *Mining Frontiers of the Far West, 1848–1880*. New York: Holt, Rinehart, and Winston, 1963.

Peterson, Richard H. *The Bonanza Kings: The Social Origins and Business Behavior of Western Mining Entrepreneurs, 1870–1900*. Lincoln: University of Nebraska Press, 1977.

———. "Conflict and Consensus: Labor Relations in Western Mining." *Journal of the West* 12 (January 1973): 1–17.

———. "The Frontier Thesis and Social Mobility on the Mining Frontier." *Pacific Historical Review* 44 (February 1975): 52–67.

———. "Philanthropic Phoebe: The Educational Charity of Phoebe Apperson Hearst." *California History* 64 (Fall 1985): 284–89, 313–15.

———. "Simeon Gannett Reed and the Bunker Hill and Sullivan: The Frustrations of a Mining Investor." *Idaho Yesterdays* 23 (Fall 1979): 2–8.

———. "Thomas F. Walsh: The Lifestyle of a Colorado Mining Magnate, 1896–1910." *Red River Valley Historical Review* 6 (Fall 1981): 53–63.

Pomeroy, Earl. "Toward a Reorientation of Western History: Continuity and Environment." *Mississippi Valley Historical Review* 41 (March 1955): 579–600.

Press Club of Salt Lake. *Men of Affairs in the State of Utah, A Newspaper Reference Work*. Salt Lake City: Press Club of Salt Lake, 1914.

Quiett, Glenn C. *They Built the West: An Epic of Rails and Cities*. New York: Cooper Square Publishers, 1965 (orig. pub. 1934).

Ratner, Sidney, ed. *New Light on the History of Great American For-*

tunes, American Millionaires of 1892 and 1902. New York: Augustus M. Kelley, Inc., 1953.

Redlich, Fritz. "The Business Leader as a 'Daimonic' Figure." *American Journal of Economics and Sociology* 20 (January–April 1953): 163–78, 289–99.

Reece, Daphne. *Historic Houses of California*. San Francisco: Chronicle Books, 1983.

Reid, J. Juan. *Colorado College: The First Century, 1874–1974*. Colorado Springs: Colorado College, 1979.

Reynolds, Clifford P., ed. *Biographical Directory of the American Congress, 1774–1961*. Washington, D.C.: Government Printing Office, 1961.

Ross, Elizabeth D. *The Kindergarten Crusade: The Establishment of Preschool Education in the United Sates*. Athens: Ohio University Press, 1976.

Rothman, David J. *Politics and Power: The United States Senate, 1869–1901*. Cambridge, Mass.: Harvard University Press, 1966.

Roy, William G. "Collecting Data on American Business Officials in the Late Nineteenth and Early Twentieth Century." *Historical Methods* 15 (Fall 1982): 143–51.

Rudd, Hynda. "Samuel Newhouse: Utah Mining Magnate and Land Developer." *Western States Jewish Historical Quarterly* 11 (July 1979): 291–307.

Ruetten, Richard. "In Re: William Andrews Clark." *Arizona and the West* 3 (Winter 1961): 170.

Schoch, Russell. "Gym Dandy." *California Monthly* 95 (October 1984): 27–29.

Schwantes, Carlos A. *Coxey's Army: An American Odyssey*. Lincoln: University of Nebraska Press, 1985.

Seligman, Ben B. *The Potentates: Business and Businessmen in American History*. New York: Dial Press, 1971.

Seligman, Germain. *Merchants of Art, 1880–1960*. New York: Appleton-Century-Crofts, 1961.

Shoebotham, H. Minar. *Anaconda, The Life of Marcus Daly, the Copper King*. Harrisburg: Stackpole Co., 1956.

Shumate, Albert. *A Visit to Rincon Hill and South Park*. San Francisco: Yerba Buena Chapter, E. Clampus Vitus, 1963.

Simon, Kate. *Fifth Avenue: A Very Social History*. New York: Harcourt, Brace and Jovanovich, 1978.

Smiley, Jerome C. *History of Denver, with Outlines of the Earlier History of the Rocky Mountain Country*. Denver: J. H. Williamson and Co., 1903 (orig. pub. 1901).

———. *Semi-Centennial History of the State of Colorado*. 2 vols. Chicago: Lewis Publishing Co., 1913.

Smith, Duane A. *Horace Tabor: His Life and the Legend*. Boulder: Colorado Associated University Press, 1973.

Smurr, John W. "The Montana 'Tax Conspiracy' of 1889." *Montana: The Magazine of Western History* 5 (Spring 1955): 46–53, and (Summer 1955): 47–56.

Spence, Clark C. "The Mining Engineers in the West." In *The American West, an Appraisal*, edited by Robert G. Ferris, 100–111. Santa Fe: Museum of New Mexico Press, 1963.

———. "Western Mining." In *Historians and the American West*, edited by Michael Malone, 96–122. Lincoln: University of Nebraska Press, 1983.

Sprague, Marshall. *Money Mountain: The Story of Cripple Creek Gold*. Boston: Little, Brown and Co., 1953.

———. *Newport in the Rockies: The Life and Good Times of Colorado Springs*. Chicago: Sage/Swallow Press Books, 1980.

Stadtman, Verne A. *The University of California, 1868–1968*. New York: McGraw-Hill, 1970.

Starr, Kevin. *Americans and the California Dream, 1850–1915*. New York: Oxford University Press, 1973.

Stevens, Otheman, ed. *Press Reference Library, Notables of the Southwest*. Los Angeles: Los Angeles Examiner, 1912.

Stewart, Robert E., Jr., and Mary F. Stewart. *Adolph Sutro; A Biography*. Berkeley: Howell-North, 1962.

Stoehr, C. Eric. *Bonanza Victorian: Architecture and Society in Colorado Mining Towns*. Albuquerque: University of New Mexico Press, 1975.

Stone, Wilbur Fisk, ed. *History of Colorado*. 4 vols. Chicago: S. J. Clark Co., 1918.

Swanberg, W. A. *Citizen Hearst: A Biography of William Randolph Hearst*. New York: Charles Scribner's Sons, 1961.

Thompson, Jacqueline. *The Very Rich Book*. New York: William Morrow and Co., 1981.

Thoresen, Timothy H. H. "Paying the Piper and Calling the Tune: The Beginnings of Academic Anthropology in California." *Journal of the History of the Behavioral Sciences* 11 (July 1975): 257–75.

Thorndike, Joseph J., Jr. *The Very Rich: A History of Wealth*. New York: American Heritage Publishing Co., Bonanza Books, 1981.

Tilton, Cecil G. *William Chapman Ralston, Courageous Builder*. Boston: Christopher Publishing House, 1935.

Toole, K. Ross, and Edward Butcher. "Timber Depredations on the Montana Public Domain, 1885–1918." *Journal of the West* 7 (July 1968): 351–62.

Turner, Frederick Jackson. *The Frontier in American History*. New York: Henry Holt and Co., 1920.

———. "The Significance of the Frontier in American History." *American Historical Association Annual Report, 1893*. Washington, D.C.: American Historical Association, 1894.

Utne, Eve Marie, ed. *Victorian Traveler's Companion*. Philadelphia: The Victorian Society in America, 1980.

Van West, Carroll. "Montana's Self-Confessed Elite: The Progressive Men of Montana." *Montana: The Magazine of Western History* 33 (Autumn 1983): 36–45.

Veblen, Thorstein. *The Theory of the Leisure Class*. New York: Modern Library, Inc., 1934 (orig. pub. 1899).

Waters, Frank J. *Midas of the Rockies: The Story of Stratton and*

Cripple Creek. Denver: Alan Swallow, 1949 (orig. pub. 1937).

Webster, George H. "El Paso Club of Colorado Springs." *Colorado Magazine* 26 (April 1949): 104–12.

Weinstein, Allen. *Prelude to Populism: Origins of the Silver Issue, 1867–1878*. New Haven: Yale University Press, 1970.

Who's Who in America. Vols. 5 and 16. Chicago: A. N. Marquis and Co., 1908–9 and 1930–31.

Who Was Who in America, 1897–1942. Chicago: A. N. Marquis Co., 1943.

Wildman, Edwin. *Famous Leaders of Industry: The Life Stories of Boys Who Have Succeeded*. Boston: The Page Co., 1920.

Williams, R. Hal. *The Democratic Party and California Politics, 1880–1896*. Stanford: Stanford University Press, 1973.

World Book Encyclopedia. 22 vols. and supplements. Chicago: World Book–Childcraft International, Inc., 1981.

Wright, Georgia Sommers. "Architect's Mind in Flux: Maybeck's *Beaux-Arts* Triumphs, Trials, and Evolution at Mills College." *The Californian* 2 (July/August 1984): 31–38.

Wright, James E. *The Politics of Populism: Dissent in Colorado*. New Haven: Yale University Press, 1974.

Wright, Louis B. *Culture on the Moving Frontier*. Bloomington: Indiana University Press, 1955; New York: Harper and Bros., 1961.

———. "The Westward Advance of the Atlantic Frontier." *Huntington Library Quarterly* 11 (May 1948): 261–75.

Wyllie, Irvin G. *The Self-Made Man in America: The Myth of Rags to Riches*. New York: The Free Press, 1966.

———. "Social Darwinism and the Businessmen." *Proceedings of the American Philosophical Society* 103, no. 5 (1959): 629–35.

Index